United Kingdom and Red Ensign Flag Administrations

THE CODE OF PRACTICE FOR SAFETY OF LARGE COMMERCIAL SAILING & MOTOR VESSELS

Large is 24 metres and over in load line length and the Code of Practice applies for yachts which are in commercial use for sport or pleasure, do not carry cargo and do not carry more than 12 passengers.

AN EXECUTIVE AGENCY OF THE DEPARTMENT OF TRANSPORT

London: The Stationery Office

Published with the permission of the Marine Safety Agency on behalf of the controller of Her Majesty's Stationery Office.

ISBN 0 11 551911 4

TABLE OF CONTENTS

1 Foreword

1.1 This Code of Practice has been developed jointly by the United Kingdom and its relevant overseas territories which have been assigned Category 1 registry by the Merchant Shipping (Categorisation of Registries of Overseas Territories) Order 1992, SI 1992 No.1736. The overseas territories assigned Category 1 (unlimited tonnage and type) registry by the order are:-

Bermuda
Cayman Islands
Isle of Man.

Where "Administration" is used in the Code, it applies to the United Kingdom of Great Britain and Northern Ireland or the overseas territory assigned Category 1 registry.

Overseas territories assigned Category 1 registry may be amended from time to time and reference should be made to the above Order and any legislation either amending or replacing it. Annex 1 records the status of overseas territories at the time the Code was published for the first time in 1997.

1.2 Vessels are required to comply with the various Merchant Shipping Regulations of the flag Administration which are relevant to the class of vessel to which they belong. Vessels in commercial use for sport or pleasure do not fall naturally into a single class and, in any case, prescribed merchant ship safety standards may be incompatible with the safety needs particular to such vessels.

This version of the Code refers to the regulations of the United Kingdom of Great Britain and Northern Ireland. Annex 3 lists documents to which the Code refers for the application of specific safety and pollution prevention standards.

1.3 The Code applies to vessels in commercial use for sport or pleasure (being pleasure vessels "engaged in trade" for the purpose of Article 5 - Exceptions - of the International Convention on Load Lines, 1966 (ICLL)) which are 24 metres in load line length and over or, if built before 21 July 1968, 150 gross tons and over according to the tonnage measurement regulations applying at that date and which do not carry cargo and do not carry more than 12 passengers.

1.4 On behalf of the overseas territories (named in 1.1) to which the International Convention on Load Lines, 1966 has been extended, the United Kingdom has notified the International Maritime Organisation of the Code and its application to pleasure vessels engaged in trade as an equivalent arrangement under the provisions of Article 8 of the International Convention on Load Lines, 1966.

1.5 The Code sets required standards of safety and pollution prevention which are appropriate to the size of the vessel. The standards applied are either set by the relevant international conventions or equivalent standards where it is not reasonable or practicable to comply with the international convention.

An Administration may consider a specific alternative equivalent standard to any standard required by the Code. Applications which justify either an alternative or exemption from a specific requirement of the Code can be made to the Administration.

1.6 The Code of Practice has been developed by an industry wide group with the express intention of setting pollution prevention and safety standards which identify with the specific needs of vessels in commercial use for sport and pleasure. The standards adopted are judged to be at least equivalent in their effect to those required by the international conventions which apply to a particular vessel.

The membership of industry wide group which developed the Code are listed in Annex 3.

1.7 Compliance with the standards required by the Code will entitle a vessel to be issued with the certification required by the international conventions applicable to the vessel, upon satisfactory completion of the corresponding surveys and inspections.

The certificates demanded by the international conventions which apply to the vessels covered by the Code are summarised in section 28.

1.8 When equipment manufactured in accordance with a recognised British, European or International standard is required by the Code, the Administration may accept existing equipment which can be shown to be of an equivalent standard and which does not constitute a risk to the ship or its crew and passengers. When such equipment is replaced, the replacement should to the standard required by the Code.

1.9 The Commission of the European Unions' general mutual recognition clause should be accepted. The clause states:-

Any requirement for goods or materials to comply with a specified standard should be satisfied by compliance with:-

.1 a relevant standard or code of practice of a national standards body or equivalent body of a Member State of the European Union; or

.2 any relevant international standard recognised for use in any Member State of the European Union; or

.3 a relevant specification acknowledged for use as a standard by a public authority of any Member State of the European Union; or

.4 traditional procedures of manufacture of a Member State of the European Union where these are the subject of a written technical description sufficiently detailed to permit assessment of the goods or materials for the use specified; or

.5 a specification sufficiently detailed to permit assessment for goods or materials of an innovative nature (or subject to innovative processes of manufacture such that they cannot comply with a recognised standard or specification) and which fulfil the purpose provided by the specified standard;

provided that the proposed standard, code of practice, specification or technical description provides, in use, equivalent levels of safety, suitability and fitness for purpose.

1.10 It is recognised that the Code may be required to be revised in the light of experience gained in its application. Section 3.4 makes provision for the Code to be revised.

1.11 It is recommended that pleasure vessels as defined in the Merchant Shipping (Vessels in Commercial Use for Sport or Pleasure) Regulations 1993, SI 1993 No.1072, as amended, comply with the standards of the Code.

2 Definitions

"Administration" with regard to this Code and the flag the vessel is entitled to fly, means, as appropriate, the Government of the United Kingdom or the overseas territory assigned Category 1 registration by the United Kingdom Merchant Shipping (Categorisation of Registries of Overseas Territories) Order currently in force or an organisation formally authorised or appointed by the Administration to act on its behalf with regard to the conduct of specified survey and/or issue of certification.

(The Merchant Shipping Regulation references in this Code are appropriate to the Government of the United Kingdom of Great Britain and Northern Ireland);

"Approved" in respect to materials or equipment means approved by the Administration or approved by an administration or organisation which is formally recognised by the Administration;

"Authorised surveyor" means a surveyor who by reason of professional qualifications, practical experience and expertise is authorised to carry out the survey required, by the Administration for the vessel;

"Auxiliary steering gear" is the equipment other than any part of the main steering gear necessary to steer the ship in the event of failure of the main steering gear but not including the tiller, quadrant or components serving the same purpose;

"Buoyant lifeline" means a line complying with the requirements of Part V of Schedule 9 of the Merchant Shipping (Life-Saving Appliances) Regulations 1986;

"Buoyant smoke signal" means a pyrotechnic signal complying with the requirements of Part IV of Schedule 8 of the Merchant Shipping (Life-Saving Appliances) Regulations 1986;

"Cargo" means an item(s) of value that is carried from one place and discharged at another place and for which either a charge or no charge is made and is not for use exclusively onboard the vessel;

"Commercial vessel" means a vessel which is registered under Part I of the Registration Act and is described in the register and on the certificate of registry as a commercial vessel and is not a pleasure vessel;

"Control stations" are those spaces in which the ship's radio or main navigating equipment or the emergency source of power is located or where the fire recording or fire control equipment is centralized;

"Date of expiry" in relation to pyrotechnics and self-activating smoke signals referred to in Schedules 8 and 9 of the Merchant Shipping (Life-Saving Appliances) Regulations 1986 means a date within 3 years from the date of manufacture of that product;

"Dead ship condition" is the condition under which the main propulsion plant, boilers and auxiliaries are not in operation due to the absence of power;

"Efficient" in relation to a fitting, piece of equipment or material means that all reasonable and practicable measures have been taken to ensure that it is suitable for the purpose for which it is intended to be used;

"Embarkation ladder" means a ladder complying with the requirements of Part V of Schedule 6 of the Merchant Shipping (Life-Saving Appliances) Regulations 1986 provided at embarkation stations to permit safe access to survival craft after launching;

"Emergency condition" is a condition under which any services needed for normal operational and habitable conditions are not in working order due to failure of the main source of electrical power;

"Emergency source of electrical power" is a source of electrical power, intended to supply the emergency switchboard in the event of failure of the supply from the main source of electrical power;

"Emergency switchboard" is a switchboard which in the event of failure of the main electrical power supply system is directly supplied by the emergency source of electrical power or the transitional source of emergency power and is intended to distribute electrical energy to the emergency services;

"EPIRB" means a satellite emergency position-indicating radio beacon, being an earth station in the mobile-satellite service, the emissions of which are intended to facilitate search and rescue operations, complying with performance standards adopted by the IMO contained in either Assembly Resolution A.763(18) or Assembly Resolution A.661(16), or any Resolution amending or replacing these from time to time and which is considered by the Secretary of State to be relevant, and is capable of:-

(a) floating free and automatically activating if the ship sinks;

(b) being manually activated; and

(c) being carried by one person;

"Existing commercial vessel" means any vessel which is registered under Part I of the Registration Act and is described in the register and on the certificate of registry as a commercial vessel, the keel of which was laid or the construction or lay up was started before the 1st October 1997;

"Float-free launching" means that method of launching a liferaft whereby the liferaft is automatically released from a sinking ship and is ready for use complying with Part VI of Schedule 4 of the Merchant Shipping (Life-Saving Appliances) Regulations 1986;

"Freeboard" has the meaning given in annex I of ICLL viz. The freeboard assigned is the distance measured vertically downwards amidships from the upper edge of the deck line to the upper edge of the related load line;

"Freeboard deck" has the meaning given in annex I of ICLL viz. The freeboard deck is normally the uppermost complete deck exposed to the weather and sea, which has permanent means of closing all openings in the weather part thereof, and below which all openings in the sides of the ship are fitted with permanent means of watertight closing.

In a ship having a discontinuous freeboard deck, the lowest line of the exposed deck and the continuation of that line parallel to the upper part of the deck is taken as the freeboard deck.

At the option of the owner and subject to the approval of the Administration, a lower deck may be designated as the freeboard deck provided it is a complete and permanent deck continuous in a fore and aft direction at least between the machinery space and peak bulkheads and continuous athwartships.

When a lower deck is designated as the freeboard deck, that part of the hull which extends above the freeboard deck is treated as a superstructure so far as concerns the application of the conditions of assignment and the calculation of freeboard. It is from this deck that the freeboard is calculated;

"Garbage" means all kinds of victual, domestic and operational waste excluding fresh fish and parts thereof, generated during the normal operation of the vessel and liable to be disposed of continuously or periodically, except sewage originating from vessels;

"General emergency alarm system" means a system complying with the requirements of Schedule 13 of the Merchant Shipping (Life-Saving Appliances) Regulations 1986;

"ICLL" means the International Convention on Load Lines, 1966, as amended;

"IMO" means the International Maritime Organisation, a specialised agency of the United Nations devoted to maritime affairs;

"Inflatable lifejacket" means a lifejacket complying with the requirements of Part II of Schedule 10 of the Merchant Shipping (Life-Saving Appliances) Regulations 1986;

"Inflated boat" means a boat complying with the requirements of Schedule 3 of the Merchant Shipping (Life-Saving Appliances) Regulations 1986 and suitable for rescuing persons in distress and for marshalling liferafts;

"Instructions for on-board maintenance" means the instructions complying with the requirements of Part II of Schedule 12 of the Merchant Shipping (Life-Saving Appliances) Regulations 1986;

"Launching appliance" means a provision complying with the requirements of Part I and the relevant requirements of Parts II, III or IV of Schedule 6 of the Merchant Shipping (Life-Saving Appliances) Regulations 1986 for safely transferring a lifeboat, rescue boat, liferaft or inflated boat respectively, from its stowed position to the water and recovery where applicable;

"Length" means 96% of the total length on a waterline of a ship at 85% of the least moulded depth measured from the top of the keel, or the length from the fore-side of the stem to the axis of the rudder stock on that waterline, if that be greater. In ships designed with a rake of keel the waterline on which this is measured shall be parallel to the designed waterline;

"Lifeboat" means a lifeboat complying with the requirements of Parts I and II of Schedule 1 of the Merchant Shipping (Life-Saving Appliances) Regulations 1986;

"Lifebuoy" means a lifebuoy complying with the requirements of Part I of Schedule 9 of the Merchant Shipping (Life-Saving Appliances) Regulations 1986;

"Lifejacket" means a lifejacket complying with the requirements of parts I or II of Schedule 10 of the Merchant Shipping (Life-Saving Appliances) Regulations 1986;

"Liferaft" means a liferaft complying with the requirements of Part I of Schedule 4 of the Merchant Shipping (Life-Saving Appliances) Regulations 1986;

"Line throwing appliance" means an appliance complying with the requirements of Part V of Schedule 8 of the Merchant Shipping (Life-Saving Appliances) Regulations 1986;

"Low flame spread" means that the surface thus described will adequately restrict the spread of flame, this being determined to the satisfaction of the Administration by an established procedure;

"Machinery spaces" are all machinery spaces of category A and all other spaces containing propelling machinery, boilers, oil fuel units, steam and internal combustion engines, generators and major electrical machinery, oil filling stations, refrigerating, stabilizing, ventilation and air conditioning machinery, and similar spaces, and trunks to such spaces;

"Machinery spaces of category A" are those spaces and trunks to such spaces which contain:

(a) internal combustion machinery used for main propulsion; or

(b) internal combustion machinery used for purposes other than main propulsion where such machinery has in the aggregate a total power output of not less than 375 kW; or

(c) any oil fired boiler or oil fuel unit;

"Main generating station" is the space in which the main source of electrical power is situated;

"Main source of electrical power" is a source intended to supply electrical power to the main switchboard for distribution to all services necessary for maintaining the ship in normal operational and habitable condition;

"Main steering gear" is the machinery, rudder actuators, steering gear power units, if any, and ancillary equipment and the means of applying torque to the rudder stock (e.g. tiller or quadrant) necessary for effecting movement of the rudder for the purpose of steering the ship under normal service conditions;

"Main switchboard" is a switchboard which is directly supplied by the main source of electrical power and is intended to distribute electrical energy to the ship's services;

"Main vertical zone" means those sections into which the hull, superstructure and deckhouses are divided by A class divisions, the mean length of which on any deck does not normally exceed 40 metres;

"MARPOL" means the International Convention for the Prevention of Pollution from Ships, 1973, as amended;

"Marine Safety Agency" means the Marine Safety Agency, an executive agency of the United Kingdom Department of Transport;

"Maximum ahead service speed" for the purpose of steering gear and rudder stock and pintle design, is the maximum contractual speed of the ship, in knots;

"Merchant Shipping Notices", "Marine Guidance Notes" and "Marine Information Notes" means Notices/Notes described as such and issued by the Department of Transport; and any reference to a particular Notice/Note includes a reference to any document amending or replacing that Notice/Note which is considered by the Secretary of State to be relevant from time to time and is specified in a Notice/Note;

"Mile" means a nautical mile of 1852 metres;

"Motor vessel" means a commercial vessel which is described in the register and on the certificate of registry as such, and which has as a sole means of propulsion either one or more power units;

"Multihull vessel" means any vessel which in any normally achievable operating trim or heel angle, has a rigid hull structure which penetrates the surface of the sea over more than one separate or discrete area;

"New vessel" means a vessel to which this Code applies, the keel of which was laid or the construction or lay up was started on or after the 1st October 1997;

"Not readily ignitable" means that the surface thus described will not continue to burn for more than 20 seconds after removal of a suitable impinging test flame;

"Operated on a commercial basis" means the operation of the ship is being financed either wholly or in part by persons or a company other than the owner, or the immediate family of the owner;

"Owner(s)/managing agent(s)" means the registered owner(s) or the owner(s) or the managing agent(s) of the registered owner(s) or the owner(s) or owner(s) ipso facto, as the case may be;

"Passenger" means any person carried in a ship except:

(a) a person employed or engaged in any capacity on board the ship on the business of the ship;

(b) a person on board the ship either in pursuance of the obligation laid upon the master to carry shipwrecked, distressed or other persons, or by reason of any circumstances that neither the master nor the owner nor the charterer (if any) could have prevented; and

(c) a child under one year of age;

"Passenger ship" means a ship carrying more than 12 passengers;

"Person" means a person over the age of one year;

"Pleasure vessel" means a vessel so defined in the Merchant Shipping (Vessels in Commercial Use) Regulations 1993;

"Position 1" means upon exposed freeboard and raised quarter decks and upon exposed superstructure decks situated forward of a point located a quarter of the ship's length from the forward perpendicular;

"Position 2" means upon exposed superstructure decks situated abaft a quarter of the ship's length from the forward perpendicular;

"Power actuating system" is the hydraulic equipment provided for supplying power to turn the rudder stock, comprising a steering gear power unit or units, together with the associated pipes and fittings, and a rudder actuator. The power actuating systems may share common mechanical components, i.e., tiller, quadrant and rudder stock, or components serving the same purpose;

"Radar transponder" means a radar transponder for use in survival craft to facilitate location of survival craft in search and rescue operations;

"Recess" means an indentation or depression in a deck and which is surrounded by the deck and has no boundary common with the shell of the vessel;

"Rescue boat" means a boat complying with the requirements of Parts I, II or III of Schedule 2 of the Merchant Shipping (Life-Saving Appliances) Regulations 1986 and designed to rescue persons in distress and for marshalling liferafts;

"Retro-reflective material" means a material which reflects in the opposite direction a beam of light directed on it and which complies with the specification laid down in Merchant Shipping Notice No.M.1444;

"Rocket parachute flare" means a pyrotechnic signal complying with the requirements of Part I of Schedule 8 of the Merchant Shipping (Life-Saving Appliances) Regulations 1986;

"Safe haven" means a harbour or shelter of any kind which affords entry, subject to prudence in the weather conditions prevailing, and protection from the force of the weather;

"Sailing vessel" means a vessel designed to carry sail, whether as a sole means of propulsion or as a supplementary means;

"Sail training vessel" means a sailing vessel which at the time it is being used is being used either:-

(a) to provide instruction in the principles of responsibility, resourcefulness, loyalty and team endeavour and to advance education in the art of seamanship; or

(b) to provide instruction in navigation and seamanship for yachtsmen;

"Self-activating smoke signal" means a signal complying with the requirements of Part IV of Schedule 9 of the Merchant Shipping (Life-Saving Appliances) Regulations 1986;

"Self-igniting light" means a light complying with the requirements of Part III of Schedule 9 of the Merchant Shipping (Life-Saving Appliances) Regulations 1986;

"Side scuttle" means an ISO standardised type of an opening hinged or non-opening round ship's window with or without deadlight (ISO 6345:1990);

"SOLAS" means the International Convention for the Safety of Life at Sea, 1974, as amended;

"SOLAS A emergency pack" means a liferaft emergency pack complying with requirements of Part IV of Schedule 4 of the Merchant Shipping (Life-Saving Appliances) Regulations 1986;

"SOLAS B emergency pack" means a liferaft emergency pack complying with the requirements of Part IV of Schedule 4 of the Merchant Shipping (Life-Saving Appliances) Regulations 1986;

"Steering gear control system" is the equipment by which orders are transmitted from the navigating bridge to the steering gear power units. Steering gear control systems comprise transmitters, receivers, hydraulic control pumps and their associated motors, motor controllers, piping and cables;

"Steering gear power unit" is:

(a) in the case of electric steering gear, an electric motor and its associated electrical equipment;

(b) in the case of electrohydraulic steering gear, an electric motor and its associated electrical equipment and connected pump;

(c) in the case of other hydraulic steering gear, a driving engine and connected pump;

"Superstructure" has the meaning given in annex I to ICLL;

"Survival craft" means a craft capable of sustaining the lives of persons in distress from the time of abandoning the ship;

"Training manual" with regard to life-saving appliances means a manual complying with the requirements of Part I of Schedule 12 of the Merchant Shipping (Life-Saving Appliances) Regulations 1986;

"Two-way VHF radiotelephone set" means a portable or a fixed VHF installation for survival craft complying with the performance adopted by the Organisation contained in Organisation Resolution A.762(18) or any Resolution amending or replacing it which is considered by the Secretary of State to be relevant from time to time;

"To sea" means beyond any partially smooth waters, or smooth waters limits which may have been designated by the Authority in which the vessel is operating. In the event that no such areas have been designated, then it means that the vessel is considered to have proceeded to sea upon leaving the immediate confined designated harbour;

"Voyage" includes an excursion;

"Waterproofed" means protected as far as is practicable from the ingress of water;

"Watertight" means capable of preventing the passage of water in any direction;

"Weather deck" means the uppermost complete weathertight deck fitted as an integral part of the vessel's structure and which is exposed to the sea and weather;

"Weathertight" has the meaning given in annex I of ICLL viz. Weathertight means that in any sea conditions water will not penetrate into the ship;

"Wheelhouse" means the control position occupied by the officer of the watch who is responsible for the safe navigation of the vessel;

"Window" means a ship's window, being any window, regardless of shape, suitable for installation aboard ships (ISO 6345:1990).

3 Application and Interpretation

3.1 Application

3.1.1 The Code applies to a motor or sailing vessel of 24 metres in load line length and over or, if built before 21 July 1968, which is of 150 tons gross tonnage and over and which, at the time, is in commercial use for sport or pleasure and carries no cargo and up to 12 passengers, provided that it is not a vessel to which either the International Code of Safety for High Speed Craft or the Code of Safety for Dynamically Supported Craft is applicable. Sail training vessels are included in this application.

3.1.2 Compliance with this Code satisfies the requirements of the Merchant Shipping (Vessels in Commercial Use for Sport or Pleasure) Regulations 1993, SI 1993 No.1072, as amended.

3.1.3 For the application of the Code, the Regulations insist that any provision of the Code expressed in the conditional (i.e. "should") shall be a requirement.

3.2 Equivalent Standards and Exemptions and Existing Vessels

3.2.1 Equivalent standards

Proposals for the application of alternative standards considered to be at least equivalent to the requirements of the Code should be submitted to the Administration for approval. Equivalence may be achieved by incorporating increased requirements to balance deficiencies and thereby achieve the overall safety standard.

3.2.2 Exemptions

Exemptions should be granted only by the Administration.

Applications for exemption should be made to the Administration and be supported by justification for the exemption.

The granting of exemptions will be limited by the extent to which international conventions allow and should be regarded as exceptional.

3.2.3 Existing vessels

3.2.3.1 In the case of an existing vessel which does not comply fully with the Code safety standards but for which the Code standards are reasonable and practicable, the Administration should give consideration to a proposal from the owner(s)/managing agent(s) to phase in requirements within an agreed time scale.

3.2.3.2 When an existing vessel does not meet the Code safety standard for a particular feature and it can be demonstrated that compliance is neither reasonable nor practicable, proposals for alternative arrangements should be submitted to the Administration for approval. In considering individual cases, the Administration should take into account the vessel's service history and any other factors which are judged to be relevant to the safety standard which can be achieved.

3.2.3.3 Generally, repairs, alterations and refurbishments should comply with the standards applicable to a new vessel.

3.3 Interpretation

Where a question of interpretation of any part of this Code arises which cannot be resolved by the Classification Society, Overseas Territory, Certifying Authority in respect of radio surveys and the owner(s)/managing agent(s) for a vessel, a decision on the interpretation may be obtained on written application to:

The Director
Standards Division
Marine Safety Agency
Spring Place
105 Commercial Road
Southampton SO15 1EG

Facsimile: National 01703 329161
International +44 1703 329161

3.4 Updating the Code

The requirements of the Code will be reviewed and, if necessary revised, by a standing committee, comprising representatives from the industry wide group which developed the Code, within five years of its coming into force through the enabling statutory legislation. Any amendments which are required before such time will be promulgated by issue of Merchant Shipping Notices.

4 Construction and Strength

4.1 General Requirements

4.1.1 All vessels should have a freeboard deck.

4.1.2 All vessels should be fitted with a weathertight weather deck throughout the length of the vessel and be of adequate strength to withstand the sea and weather conditions likely to be encountered in the declared area(s) of operation.

4.1.3 The declared area(s) of operation and any other conditions which restrict the use of the vessel at sea should be recorded on the load line certificate issued to the vessel.

4.1.4 The choice of hull construction material affects fire protection requirements, for which reference should be made to section 14.

4.2 Structural Strength

4.2.1 New vessels

4.2.1.1 New vessels will be considered to be of adequate strength if built under survey and are certificated to be in accordance with hull certification standards set by any of the following organisations:

British Technical Committee of American Bureau of Shipping
British Committee of Bureau Veritas
British Committee of Det Norske Veritas
British Committee of Germanischer Lloyd
British Committee of Registro Italiano Navale
Lloyd's Register of Shipping

A classification certificate should be provided.

4.2.1.2 New vessels not built in accordance with 4.2.1.1 may be specially considered provided full information, including calculations, drawings and details of materials is presented to the Administration for approval and, subject to a satisfactory survey by an Authorised Surveyor.

A certificate of survey should be provided.

(See section 28 regarding survey during construction.)

4.2.1.3 New vessels classed by one of the Classification Societies listed in 4.2.1.1 after construction has been completed in accordance with the standards of the Society, will be accepted as being of adequate strength for the service conditions covered by the classification notation.

A classification certificate should be provided.

4.2.2 Existing vessels

Existing vessels will be considered to be of adequate structural strength if they are in good repair and were:

.1 Built to the standards defined by 4.2.1.1 for new vessels and remain in class, or

.2 Built to the standards defined by 4.2.1.1 for new vessels and, where no longer in class, are subjected to a full structural survey by an Authorised Surveyor to determine its condition.

A certificate of survey should be provided.

.3 Not built in accordance with 4.2.2.1 but where full information, including calculations, drawings and details of materials has been presented and accepted by the Administration and, subject to a satisfactory survey by an Authorised Surveyor.

A certificate of survey should be provided.

4.3 Recesses

4.3.1 Any recess in the weathertight weather deck should be of weathertight construction and should be self draining under all normal conditions of heel and trim of the vessel.

A swimming pool open to the elements should be treated as a recess.

4.3.2 The means of drainage provided should be capable of efficient operation when the vessel is heeled to an angle of 10° in the case of a motor vessel (see 10A.2), and 30° in the case of a sailing vessel.

The drainage arrangements should have the capability of draining the recess (when fully charged with water) within 3 minutes when the vessel is upright and at the load line draught. Means should be provided to prevent the backflow of sea water into the recess.

4.3.3 When it is not practical to provide drainage which meets the requirements of 4.3.2, alternative safety measures may be proposed for approval by the Administration.

4.4 Watertight Bulkheads

Section 11 of the Code deals with subdivision and damage stability requirements which will determine the number and positioning of watertight bulkheads defined below.

4.4.1 New vessels

Watertight bulkheads should be fitted in accordance with the following requirements:

.1 The strength of watertight bulkheads should be in accordance with the requirements of one of the Classification Societies listed in 4.2.1.1.

.2 Generally, openings in watertight bulkheads should comply with the standards required for passenger vessels. The standards are defined in SOLAS regulations II-1.

.3 In exceptional circumstances the Administration may accept hinged watertight doors in lieu of those required by SOLAS regulations II-1.

4.4.2 Existing vessels

4.4.2.1 Watertight bulkheads in existing vessels should comply with the requirements of 4.4.1 as far as it is reasonable and practicable to do so.

4.4.2.2 In individual cases, when the Administration considers that the requirements of 4.4.1 cannot be met, the Administration may consider a justification for exemption from the specified requirements.

4.4.2.3 In considering an individual case, the Administration will take into account the vessels past performance in service and the declared area(s) of operation and any other conditions which restrict the use of the vessel at sea, which will be recorded on the load line certificate issued to the vessel. (See 4.1.3.)

4.5 Enclosed Compartments within the Hull and below the Freeboard Deck provided with Access through Openings in the Hull

4.5.1 Compartment(s) below the freeboard deck, provided for recreational purposes, oil fuelling/fresh water reception or other purposes to do with the business of the vessel and having access openings in the hull, should be bounded by watertight divisions without any opening (ie doors, manholes, ventilation ducts or any other opening), separating the compartment(s) from any other compartment below the freeboard deck.

4.5.2 Openings in the hull should comply with SOLAS regulation II-1/25-10 - External openings in cargo ships.

4.6 Rigging on Sailing Vessels

4.6.1 General

4.6.1.1 The overall sail area and spar weights and dimensions should be as documented in the vessels's stability information booklet. Any rigging modifications that increase the overall sail area, or the weight/dimensions of the rig aloft, must be accompanied by an approved updating of the stability information booklet.

4.6.1.2 The condition of the rig should be monitored in accordance with a planned maintenance schedule. The schedule should include, in particular, regular monitoring of all the gear associated with safe work aloft and on the bowsprit (see 22.3).

4.6.2 Masts and spars

4.6.2.1 Dimensions and construction materials of masts and spars should be in accordance with the recommendations of one of the Classification Societies listed in 4.2.1.1 or a recognised national or international standard.

4.6.2.2 The associated structure for masts and spars (including fittings, decks and floors) should be constructed to absorb the forces involved.

4.6.3 Running and standing rigging

4.6.3.1 Wire rope used for standing rigging (stays or shrouds) should not be flexible wire rope (fibre rope core).

4.6.3.2 The strength of all blocks, shackles, rigging screws, cleats and associated fittings and attachment points should exceed the breaking strain of the associated running or standing rigging.

4.6.3.3 Chainplates for standing rigging should be constructed to support and absorb the forces involved. Only one shroud or stay should load an individual attachment point, unless the design specifically allows for more.

4.6.4 Sails

4.6.4.1 Adequate means of reefing or shortening sail should be provided.

4.6.4.2 Vessels only engaged in short day sailing need not carry storm canvas.

4.6.4.3 All other vessels should either be provided with separate storm sails or have specific sails designated and constructed to act as storm canvas. (The latter option is standard practice in large vessels.)

5 Weathertight Integrity

For new vessels and existing vessels, the standards for achieving weathertight integrity should comply with ICLL as far as it is reasonable and practicable.

In any case the intention should be to achieve a standard of safety which is at least equivalent to the standard of ICLL.

In individual cases, when the Administration considers that the requirements of ICLL or the Code cannot be met, the Administration may consider and approve alternative arrangements to achieve adequate safety standards.

For an existing vessel, the Administration should take into account the vessel's past performance in service and the declared area(s) of operation and any other conditions which restrict the use of the vessel at sea. Conditions which restrict the use of the vessel at sea should be recorded on the load line certificate issued to the vessel. (See 4.1.3.)

The following are illustrative standards for weathertight integrity which should be applied:

5.1 Hatchways and Skylight Hatches

5.1.1 General requirements

5.1.1.1 All openings leading to spaces below the weather deck not capable of being closed weathertight, must be enclosed within either an enclosed superstructure or a weathertight deckhouse of adequate strength.

5.1.1.2 All exposed hatchways which give access to spaces below the weathertight weather deck are to be of substantial weathertight construction and provided with efficient means of closure. Weathertight hatch covers should be permanently attached to the vessel and provided with adequate arrangements for securing the hatch closed.

5.1.1.3 Hatches which are to be used for escape purposes should be provided with covers which are capable of being opened from both sides. An escape hatch should be readily identified and easy and safe to use, having due regard to its position and access to and from the hatch.

5.1.2 Hatchways which are open at sea

In general, hatches should be kept closed at sea. However, hatchways which may be kept open for lengthy periods are to be kept as small as practicable (a maximum of 1 square metre in clear area), located on the centreline of the vessel, and fitted with coamings of at least 300mm in height. Covers of hatchways are to be permanently attached to the hatch coamings and, where hinged, the hinges are to be located on the forward side.

5.2 Doorways and Companionways

5.2.1 Doorways located above the weather deck

5.2.1.1 Exposed doors in deckhouses and superstructures which given access to spaces below the weather deck, are to be weathertight and door openings should have coaming heights of at least:

600mm when the door is in the forward quarter length of the vessel and used when the vessel is at sea;

300mm when the door is in an exposed forward facing location aft of the forward quarter length; or

150mm above the surface of the deck when the door is in a protected location aft of the forward quarter length.

5.2.1.2 Weathertight doors should be arranged to open outwards and when located in a house side, be hinged at the forward edge. Alternative closing arrangements will be considered providing it can be demonstrated that the efficiency of the closing arrangements and their ability to prevent the ingress of water will not impair the safety of the vessel.

5.2.1.3 An access door leading directly to the engine room from the weather deck should be fitted with a coaming of height 600mm if in position 1 and 380mm if in position 2.

5.2.1.4 Coaming height, construction and securing standards for weathertight doors which are provided for use only when the vessel is in port or at anchor in calm sheltered waters and are locked closed when the vessel is at sea, may be considered individually.

5.2.2 Companion hatch openings

5.2.2.1 Companionway hatch openings which give access to spaces below the weather deck should be fitted with a coaming the top of which is at least 300mm above the deck.

5.2.2.2 Washboards may be used to close the vertical opening. When washboards are used, they should be so arranged and fitted that they will not be dislodged readily. Whilst stowed, provisions are to be made to ensure that they are retained in a secure location.

5.2.2.3 The maximum breadth of an opening in a companion hatch should not exceed 1 metre.

5.3 Skylights

5.3.1 All skylights should be of efficient weathertight construction and should be located on or as near to the centreline of the vessel as practicable.

5.3.2 If they are of the opening type they should be provided with efficient means whereby they can be secured in the closed position.

5.3.3 Skylights which are provided as a means of escape should be operable from both sides. An escape skylight should be readily identified and easy and safe to use, having due regard to its position and access to and from the skylight.

5.3.4 The skylight glazing material and its method of securing within the frame should meet the appropriate marine standards defined in equivalent British, European, national or international standards.

A minimum of one portable cover for each size of glazed opening should be provided which can be accessed rapidly and efficiently secured in the event of a breakage of the skylight.

5.4 Side Scuttles

5.4.1 Side scuttles should be of an approved type. They should be of strength appropriate to location in the vessel and meet the appropriate marine standards defined in equivalent British, European, national or international standards. With regard to structural fire protection in new vessels, the requirements for the construction of certain side scuttles should meet the requirements of 14B.3.10.

5.4.2 In general, all side scuttles fitted in locations protecting openings to spaces below the weather deck or fitted in the hull of the vessel should be provided with a deadlight which is to be permanently attached and is capable of securing the opening watertight in the event of a breakage of the scuttle glazing. Proposals to fit side scuttles with portable deadlights will be subject to special consideration and approval by the Administration, having regard for the location of the side scuttles and ready availability of deadlights to be fitted. Consideration should be given to the provision of operational instructions to the Master as to when deadlights must be applied to side scuttles.

5.4.3 Side scuttles fitted in the hull of the vessel below the level of the freeboard deck should be either non-opening or of a non readily opening type, have a glazed diameter of not more than 450mm and be in accordance with a standard recognised by the Administration. The sill height of the side scuttles should be at least 500mm or 2.5% of the breadth of the vessel, whichever is the greater, above the all seasons load line assigned to the vessel. Scuttles of the non readily opening type must be secured closed when the vessel is in navigation.

5.4.4 Side scuttles should not be fitted in the hull in way of the machinery space.

5.5 Windows

5.5.1 Windows should be of an approved type. They should be of strength appropriate to location in the vessel and meet the appropriate marine standards defined in equivalent British, European, national or international standards. With regard to structural fire protection in new vessels, the requirements for the construction of certain windows should meet the requirements of 14B.3.10.

5.5.2 In general, windows fitted in superstructures or weathertight deckhouses are to be substantially framed and efficiently secured to the structure. The glass is to be of the toughened safety glass type.

Safety standards relating to the provision of large glass doors or windows fitted in the aft end of a superstructure or weathertight deckhouse will be considered on an individual basis by the Administration.

5.5.3 In general, windows should not be fitted in the main hull below the level of the freeboard deck. Proposals to fit windows in the main hull below the level of the freeboard deck will be subject to special consideration and approval by the Administration, having regard for the location and strength of the windows and their supporting structure and, the availability of strong protective covers for the windows. One item of the special consideration should be operational instructions to the Master as to when the strong protective covers must be applied to windows.

5.5.4 Storm shutters are required for all windows in the front and sides of first tier and front windows of the second tier of superstructures or weathertight deckhouses above the freeboard deck. When storm shutters are interchangeable port and starboard, a minimum of 50% of each size should be provided.

5.5.5 Windows to the navigating position should not be of either polarised or tinted glass. (See 18.2.2.)

5.6 Ventilators and Exhausts

5.6.1 Adequate ventilation is to be provided throughout the vessel. The accommodation is to be protected from the entry of gas and/or vapour fumes from machinery, exhaust and fuel systems.

5.6.2 Ventilators are to be of efficient construction and provided with permanently attached means of weathertight closure. Generally, ventilators serving any space below the freeboard deck or an enclosed superstructure should have a coaming of minimum height:

900mm in the forward quarter length of the vessel; and
760mm elsewhere.

5.6.3 Ventilators should be kept as far inboard as practicable and the height above the deck of the ventilator opening should be sufficient to prevent the ingress of water when the vessel heels.

5.6.4 The ventilation of spaces such as the machinery space, which must remain open, requires special attention with regard to the location and height of the ventilation openings above the deck, taking into account the effect of downflooding angle on stability standard. (See section 11.)

The means of closure of ventilators serving the machinery space should be selected with regard to the fire protection and extinguishing arrangements provided in the machinery space.

5.6.5 Engine exhaust outlets which penetrate the hull below the freeboard deck should be provided with means to prevent backflooding into the hull through a damaged exhaust system.

5.7 Air Pipes

5.7.1 Air pipes serving fuel and other tanks should be of efficient construction and provided with permanently attached means of weathertight closure. Means of closure may be omitted if it can be shown that the open end of an air pipe is afforded adequate protection by other structure(s) which will prevent the ingress of water.

5.7.2 Where located on the weather deck, air pipes should be kept as far inboard as practicable and be fitted with a coaming of sufficient height to prevent inadvertent flooding. Generally, air pipes to tanks should have a minimum coaming height of:

760mm when sited on the freeboard deck; and
450mm elsewhere.

5.7.3 Air pipes to fuel tanks should terminate at a height of not less than 760mm above either, the top of the filler pipe for a gravity filling tank or, the top of the overflow tank for a pressure filling tank.

5.8 Scuppers, Sea Inlets and Discharges

The standards of ICLL should be applied to every discharge led through the shell of the vessel as far as it is reasonable and practicable to do so, and in any case, all sea inlet and overboard discharges should be provided with efficient shut-off valves arranged in positions where they are readily accessible at all times.

5.9 Materials for Valves and Associated Piping

5.9.1 Valves which are fitted below the waterline should be of steel, bronze or other material having a similar resistance to impact and fire.

5.9.2 The associated piping should, in areas as indicated above, be of steel, bronze, copper or other equivalent material.

5.9.3 Where the use of plastic piping is proposed, it will be considered on an individual basis and full details of the type of piping, its intended location, and use, should be submitted to the Administration for approval. The Administration may require tests to be carried out on the plastic piping as necessary, to give approval to its use.

5.9.4 The use of flexible piping in any situation should be kept to a minimum compatible with the essential reason for its use. Flexible piping and the means of joining it to its associated hard piping system should be approved as fit for the purpose by the Administration.

6 Water Freeing Arrangements

6.1 For new vessels and existing vessels, the standards for water freeing arrangements should comply with ICLL as far as it is reasonable and practicable to do so.

In any case the intention should be to achieve a standard of safety which is at least equivalent to the standard of ICLL.

6.2 In individual cases, when the Administration considers that the requirements of ICLL cannot be met, the Administration may consider and approve alternative arrangements to achieve adequate safety standards.

In considering an individual case, the Administration will take into account the vessels past performance in service and the declared area(s) of operation and any other conditions which restrict the use of the vessel at sea which will be recorded on the load line certificate issued to the vessel. (See section 4.1.)

6.3 Section 4.3 sets requirements specific to the drainage of recesses.

7A Machinery - Vessels of less than 50 metres in length and under 500 GT

7A.1 General Requirements

7A.1.1 The machinery and its installation should, in general, meet with the requirements of one of the Classification Societies named in section 4.2.1.1. Either the vessel should be in class or a certificate of compliance issued by one of the Societies should be provided to the Administration. For existing and new vessels which operate with periodically unattended machinery spaces, the machinery and its installation should meet the standards of SOLAS regulations II-1/Part E - Additional requirements for periodically unattended machinery spaces, so far as it is reasonable and practicable to do so.

7A.1.2 The requirements for main propulsion are based upon the installation of diesel powered units. When other types of main propulsion are proposed, the arrangements and installation should be specially considered.

7A.1.3 Notwithstanding the requirements of paragraph 7A.1.1, in a fuel supply system to an engine unit, when a flexible section of piping is introduced, connections should be of a screw type or equivalent approved type. Flexible pipes should be fire resistant/metal reinforced or otherwise protected from fire. Materials and fittings should be of a suitable recognised national or international standard. In the case of an existing vessel fitted with a diesel engine in which the installation of a flexible section of piping does not immediately meet the requirements, the requirements should be met when existing fittings in the fuel supply system are replaced.

7A.2 Installation

7A.2.1 Notwithstanding the requirements referred to in 7A.1, the machinery, fuel tanks and associated piping systems and fittings should be of a design and construction adequate for the service for which they are intended, and should be so installed and protected as to reduce to a minimum any danger to persons during normal movement about the vessel, due regard being made to moving parts, hot surfaces, and other hazards.

7A.2.2 Means should be provided to isolate any source of fuel which may feed a fire in an engine space fire situation. A fuel shut-off valve(s) should be provided which is capable of being closed from a position outside the engine space. The valve(s) should be fitted as close as possible to the fuel tank(s).

7A.2.3 When a glass fuel level gauge is fitted it should be of the "flat glass" type with self closing valves between the gauge and the tank.

7B Machinery - Vessels of 50 metres in length and over or 500 GT and over

For existing and new vessels, the machinery and its installation should meet the standards of SOLAS regulations II-1/Part C - Machinery installations and II-1/Part E - Additional requirements for periodically unattended machinery spaces (when appropriate), so far as it is reasonable and practicable to do so.

In any case the intention should be to achieve a standard of safety which is at least equivalent to the standard of SOLAS. Equivalence may be achieved by incorporating increased requirements to balance deficiencies and thereby achieve the required overall standard.

8A Electrical Installations - Vessels of less than 50 metres in length and under 500 GT

8A.1 General Requirements

The electrical equipment and its installation should, in general, meet with the requirements of one of the Classification Societies named in 4.2.1.1. Either the vessel should be in class or a certificate of compliance, issued by one of the Societies, should be provided to the Administration.

8A.2 Installation

8A.2.1 Particular attention should be paid to the provision of overload and short circuit protection of all circuits, except engine starting circuits supplied from batteries.

8A.2.2 Electrical devices working in potentially hazardous areas into which petroleum vapour or other hydrocarbon gas may leak, should be provided with protection against the risk of igniting the gas.

8A.3 Emergency Lighting

8A.3.1 Lighting circuits should be distributed through the spaces so that a total blackout cannot occur due to failure of a single protective device.

8A.3.2 An emergency source of lighting should be provided which should be independent of the general lighting system and sufficient to enable persons to make their way up to the open deck and evacuate the vessel if necessary.

8A.4 Batteries

Batteries of a type suitable for marine use and not liable to leakage should be used.

8A.5 Battery Stowage

Areas in which batteries are stowed should be provided with adequate ventilation to prevent an accumulation of gas which is emitted from batteries of all types.

8A.6 Flammable Gases

Reference should be made to Section 14.1.5.

8B Electrical Installations - Vessels of 50 metres in length or over or 500 GT and over

8B.1 For existing and new vessels, the electrical equipment and its installation should meet the standards of SOLAS regulations II-1/Part D - Electrical installations and II-1/Part E - Additional requirements for periodically unattended machinery spaces (when appropriate), so far as it is reasonable and practicable to do so.

In any case, the intention should be to achieve a standard of safety which is at least equivalent to the standard of SOLAS. Equivalence may be achieved by incorporating increased requirements to balance deficiencies and thereby achieve the required overall standard.

8B.2 Flammable Gases

Reference should be made to Section 14.1.5.

9A Steering Gear - Vessels of less than 50 metres in length and under 500 GT

9A.1 General Requirements

The steering gear and its installation should, in general, meet with the requirements of one of the Classification Societies named in 4.2.1.1. Either the vessel should be in class, or a certificate of compliance issued by one of the Societies should be provided to the Administration.

In the event that the above requirements cannot be met on an existing vessel, the Administration may be requested to consider and approve alternative arrangements to achieve adequate safety standards.

9A.2 For rudder steering systems, the steering gear should be capable of turning the rudder from 35° on one side to 35° on the other side at the maximum ahead service speed of the vessel. When appropriate to the safe steering of the vessel, the steering gear should be power operated in accordance with the requirements of the Administration.

9A.3 When the steering gear is fitted with remote control, arrangements should be made for emergency steering in the event of a failure of such control.

9B Steering Gear - Vessels of 50 metres in length and over or 500 GT and over

For existing and new vessels, the steering gear and its installation should meet the standards of SOLAS regulations II-1/Part C - Machinery installations, so far as it is reasonable and practicable to do so.

In any case, the intention should be to achieve a standard of safety which is at least equivalent to the standard of SOLAS. Equivalence may be achieved by incorporating increased requirements to balance deficiencies and thereby achieve the required overall standard.

10A Bilge Pumping - Vessels of less than 50 metres in length and under 500 GT

10A.1 General Requirements

The bilge pumping equipment and its installation should, in general, meet with the requirements of one of the Classification Societies named in 4.2.1.1. Either the vessel should be in class or a certificate of compliance issued by one of the Societies should be provided to the Administration.

In the event that the above requirements cannot be met on an existing vessel, the Administration may be requested to consider alternative arrangements to achieve adequate safety standards.

10A.2 All vessels should be provided with at least two independently powered pumps and suction pipes so arranged that any compartment can be effectively drained when the vessel is heeled to an angle of 10°.

10A.3 Pumps provided should be situated in not less than two separate spaces.

10A.4 Each bilge pump suction line should be fitted with an efficient strum box.

10A.5 In the case of a vessel where the propulsion machinery space may be unmanned at any time, a bilge level alarm should be fitted. The alarm should provide an audible and visual warning in the Master's cabin and in the wheelhouse. The audible and visual alarm may be accepted elsewhere if it is considered that such a location may be more appropriate.

10A.6 Pumping and piping arrangements for bilges into which fuel or other oils of similar or higher fire risk could collect, under either normal or fault conditions, should be kept clear of accommodation spaces and separate from accommodation bilge systems. Bilge level alarms meeting the requirements of 10A.6 should be fitted to all such bilges in spaces which are unmanned at any time.

10B Bilge Pumping - Vessels of 50 metres in length and over or 500 GT and over

For existing and new vessels, the bilge pumping and its installation should meet the passenger vessel standards of SOLAS regulations II-1/Part B - Subdivision and stability, so far as it is reasonable and practicable to do so.

In any case, the intention should be to achieve a standard of safety which is at least equivalent to the standard of SOLAS. Equivalence may be achieved by incorporating increased requirements to balance deficiencies and thereby achieve the required overall standard.

N.B. At present, IMO is considering proposals to harmonise the standards for bilge pumping on passenger and cargo vessels.

11 Stability

11.1 General

11.1.1 This section deals with the standards which should be met for both intact and damaged stability.

11.1.2 An intact stability standard proposed for assessment of a vessel type which is not covered by the standards defined in the Code should be submitted to the Administration for approval at the earliest opportunity.

11.1.3 When an existing vessel either fails to meet the stability standards applied to a new vessel or has up-to-date stability information which complies with a different but defined standard, the Administration should consider the stability standard of the vessel as a special case and take into account its recorded history of safe operation.

11.2 Intact Stability Standards

11.2.1 New motor vessels

11.2.1.1 Monohulls

The curves of statical stability for seagoing conditions should meet the following criteria:-

.1 the area under the righting lever curve (GZ curve) should not be less than 0.055 metre-radians up to 30° angle of heel and not less than 0.09 metre-radians up to 40° angle of heel, or the angle of downflooding, if this angle is less;

.2 the area under the GZ curve between the angles of heel of 30° and 40° or between 30° and the angle of downflooding if this is less than 40°, should not be less than 0.03 metre-radians;

.3 the righting lever (GZ) should be at least 0.20 metres at an angle of heel equal to or greater than 30°;

.4 the maximum GZ should occur at an angle of heel of preferably exceeding 30° but not less than 25°;

.5 after correction for free surface effects, the initial metacentric height (GM) should not be less than 0.15 metres; and

.6 in the event that the vessels intact stability standard fails to comply with the criteria defined in .1 to .5 the Administration may be consulted for the purpose of specifying alternative but equivalent criteria which should be achieved.

11.2.1.2 Multihulls

The curves of statical stability for seagoing conditions should meet the following criteria:-

.1 the area under the righting lever curve (GZ curve) should not be less than 0.075 metre-radians up to an angle of 20° when the maximum righting lever (GZ) occurs at 20° and, not less than 0.055 metre-radians up to an angle of 30° when the maximum righting lever (GZ) occurs at 30° or above. When the maximum righting lever (GZ) occurs at angles between 20° and 30° the corresponding requisite area under the righting lever curve (GZ curve) should be determined by linear interpretation by the formula:-

Area to maximum righting lever, at θ° = {0.055 + 0.002(30 - θ) metre radians;

.2 the area under the GZ curve between the angles of heel of 30° and 40° or between 30° and the angle of downflooding if this is less than 40°, should not be less than 0.03 metre-radians;

.3 the righting lever (GZ) should be at least 0.20 metres at an angle of heel where it reaches its maximum;

.4 the maximum GZ should occur at an angle of heel not less than 20°;

.5 after correction for free surface effects, the initial metacentric height (GM) should not be less than 0.15 metres; and

.6 if the maximum righting lever (GZ) occurs at an angle of less than 20° approval of the stability should be considered by the Administration as a special case.

11.2.2 Existing motor vessels

11.2.2.1 The standard of stability required to be achieved by an existing vessel is generally to be as required for a new vessel.

11.2.2.2 Unless a vessel is provided with stability information which is approved to a standard recognised by the Administration and relevant to the vessel in its present condition, the vessel should be treated as if it is a new vessel.

11.2.3 New sailing vessels

11.2.3.1 Monohulls

Requirements for a new vessel are:-

.1 Curves of statical stability (GZ curves) for at least the Loaded Departure with 100% consumables and the Loaded Arrival with 10% consumables should be produced.

.2 Generally, the GZ curve required by .1 should have a positive range of not less than 90°. A positive range of less than 90° may be considered but subject to the imposition of operational limitation(s).

.3 In addition to the requirements of .2, the angle of steady heel should be greater than 15° (see figure). The angle of steady heel is obtained from the intersection of a "derived wind heeling lever" curve with the GZ curve required by .1.

In the figure:-

'dwhl' = the "derived wind heeling lever" at any angle $\theta°$.
= $0.5 \times WLO \times Cos^{1.3}\theta$

where $WLO = \dfrac{GZf}{Cos^{1.3}\theta f}$

Noting that:-

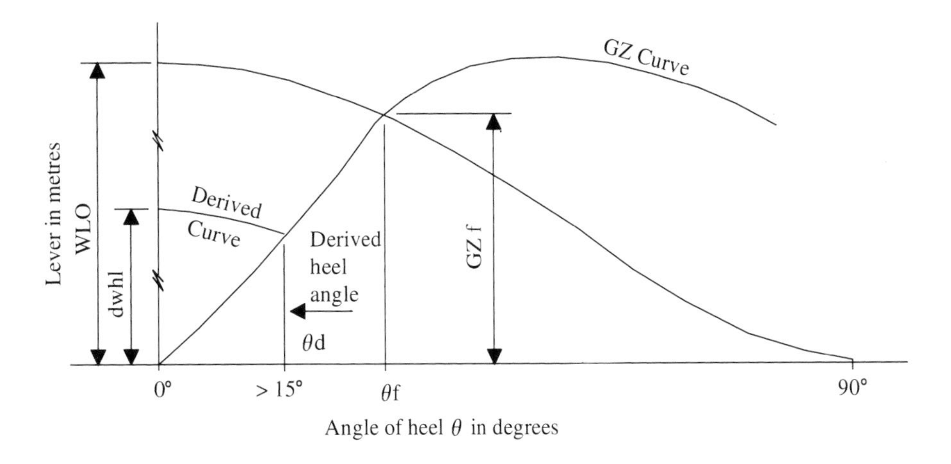

WLO is the magnitude of the actual wind heeling lever at 0° which would cause the vessel to heel to the 'down flooding angle' (θf) or 60° whichever is least.

Gzf is the lever of the vessel's GZ at the 'down flooding angle' (θf) or 60° whichever is least.

θd is the angle at which the "derived wind heeling lever" curve intersects the GZ curve. (If θd is less than 15° the vessel will be considered as having insufficient stability for the purpose of the Code).

θf the 'down-flooding angle' is deemed to occur when openings having an aggregate area, in square metres, greater than:-

$$\frac{\text{vessel's displacement in tonnes,}}{1500}$$

are immersed.

Moreover, it is the angle at which the lower edge of the actual opening which results in critical flooding becomes immersed. All regularly used openings for access and for ventilation should be considered when determining the down-flooding angle. No opening regardless of size which may lead to progressive flooding should be immersed at an angle of heel of less than 40°. Air pipes to tanks can, however, be disregarded.

If as a result of immersion of openings in a deckhouse a vessel cannot meet the required standard those deckhouse openings may be ignored and the openings in the weather deck used instead to determine θf. In such cases the GZ curve should be derived without the benefit of the buoyancy of the deckhouse.

It might be noted that provided the vessel complies with the requirements of 11.2.3.1.1, 11.2.3.1.2 and 11.2.3.1.3 and it is sailed with an angle of heel which is no greater than the 'derived angle of heel', it should be capable of withstanding a wind gust equal to 1.4 times the actual wind velocity (i.e. twice the actual wind pressure) without immersing the 'down flooding openings', or heeling to an angle greater than 60°.

11.2.3.2 Multihulls

Reference should be made to para. 11.1.2. when new multihull sailing vessels are to be certificated in accordance with this Code.

11.2.4 Existing sailing vessels

11.2.4.1 Existing monohull sailing vessels

.1 When existing valid stability information has been approved by the Marine Safety Agency or the Marine Directorate of the Department of Transport, under existing criteria, published in October 1987, this will continue to be acceptable subject to the following:-

(a) verifiable evidence is available which proves that the approved stability information is valid; or

(b) the owner/managing agent elects to re-submit a vessel for stability approval based on the new criteria.

.2 An existing monohull sailing vessel which does not comply with .1 should comply with 11.2.3.1 for a new monohull sailing vessel.

11.3 Damaged Stability

11.3.1 New vessels

11.3.1.1 The watertight bulkheads of the vessel should be so arranged that minor hull damage that results in the free flooding of any one compartment, will cause the vessel to float at a waterline which is not less than 75mm below the weather deck at any point.

11.3.1.2 Minor damage should be assumed to occur anywhere in the length of the vessel, but not on a watertight bulkhead.

11.3.1.3 Standard permeabilities should be used in this assessment, as follows:-

Space	Percentage Permeability
Stores	60
Stores but not a substantial quantity thereof	95
Accommodation	95
Machinery	85
Liquids	95 or 0 whichever results in the more onerous requirements.

11.3.1.4 In the damaged condition, considered in 11.3.1.1, the residual stability should be such that any angle of equilibrium does not exceed 7° from the upright, the resulting righting lever curve has a range to the downflooding angle of at least 15° beyond any angle of equilibrium, the maximum righting lever within that range is not less than 100mm and the area under the curve is not less than 0.015 metre radians.

11.3.1.5 For any vessels of 100 metres in length and over for which construction is started on or after 1 October 1997 and any vessels of 80 metres in length and over but less than 100 metres for which construction is started on or after 1 July 1998, application should be made to the Administration at the earliest possible stage in order to agree the damage stability requirements applicable to that vessel.

11.3.1.6 A vessel of 85 metres and above to which 13.2.1.4 - Equipment Carriage Requirements, Lifeboats - applies, should meet a 2 compartment standard of subdivision, calculated using the deterministic standard for subdivision.

11.3.2 Existing vessels

11.3.2.1 An existing vessel should be assessed according to the requirements of 11.3.1 for a new vessel. A summary of the findings should be submitted to the Administration.

11.3.2.2 When an existing vessel does not meet a standard which is required for a new vessel, the summary of findings should include a review of the consequences of overcoming the deficiency, including compensatory measures which exist or are proposed.

11.3.2.3 The Administration should accept an existing vessel on the basis of assessment made. The Administration should consider the application of operational limitations as compensation for deficiencies which cannot be overcome reasonably.

11.4 Elements of Stability

11.4.1 The lightship weight, vertical centre of gravity (KG) and longitudinal centre of gravity (LCG) of a vessel should be determined from the results of an inclining experiment.

11.4.2 An inclining experiment should be conducted in accordance with a detailed standard which is approved by the Administration and, in the presence of an authorised surveyor.

11.4.3 The report of the inclining experiment and the lightship particulars derived should be approved by the Administration prior to its use in stability calculations.

At the discretion of the owner(s)/managing agent(s) and prior to approval of the lightship particulars by the Administration, a margin for safety may be applied to the lightship weight and KG calculated after the inclining experiment.

Such a margin should be clearly identified and recorded in the stability booklet.

A formal record should be kept in the stability booklet of alterations or modifications to the vessel for which the effects on lightship weight and vertical centres of gravity are offset against the margin.

11.4.4 When sister vessels are built at the same shipyard, the Administration may accept a lightweight check on subsequent vessels to corroborate the results of the inclining experiment conducted on the lead vessel of the class.

11.5 Stability Documents

11.5.1 A vessel should be provided with a stability information booklet for the Master which is approved by the Administration.

11.5.2 The content, form and presentation of information contained in the stability information booklet should be based on the model booklet for the vessel type (motor or sailing) published by/for the Administration.

11.5.3 A vessel with previously approved stability information which undergoes a major refit or alterations should be subjected to a complete reassessment of stability and provided with newly approved stability information.

A major refit or major alteration is one which results in either a change in the lightship weight of 2% and above and/or the longitudinal centre of gravity of 1% and above (measured from the aft perpendicular) and/or the calculated vertical centre of gravity rises by 0.25% and above (measured from the keel).

11.5.4 Sailing vessels should have mounted in a suitable position for the ready reference of the crew a copy of the Curves of Maximum Steady Heel Angle to Prevent Downflooding in Squalls. This should be a direct copy taken from that contained in the approved stability booklet.

12 Freeboard

12.1 General

12.1.1 The freeboard for the vessel and its marking should be approved by the Assigning Authority for the assignment of freeboard and issue of the International Load Line Certificate (1966).

12.1.2 Vessels should comply with ICLL for the assignment of a greater than minimum freeboard mark which corresponds to the deepest loading condition included in the stability information booklet for the vessel.

12.1.3 The freeboard assigned should be compatible with the strength of hull structure and fittings, intact and damage stability requirements for the vessel.

12.1.4 The Assigning Authority should provide the owner(s)/managing agent(s) of the vessel with a copy of the particulars of the freeboard assigned and a copy of the record of particulars relating to the conditions of assignment.

12.2 Freeboard Mark and Loading

12.2.1 The freeboard mark applied should be an all seasons mark positioned port and starboard at amidships in the load line length. The mark should be permanent and be of contrasting colour to the hull of the vessel in way of the mark.

12.2.2 The fresh water freeboard allowance should be obtained by deducting from the all seasons freeboard assigned, the quantity of 1/48th of the all seasons draught of the ship at amidships.

12.2.3 A vessel should not operate in any condition which will result in its appropriate freeboard marks being submerged when it is at rest and upright in calm water.

12.3 Datum Draught Marks

12.3.1 Datum draught marks should be provided at the bow and stern, port and starboard, and be adequate in number for assessing the condition and trim of the vessel.

12.3.2 The marks should be permanent and easily read but need not be of contrasting colour to the hull. The marks need not be at more than one draught at each position but should be above and within 1000mm of the deepest load waterline.

12.3.3 The draught to which marks relate should be indicated either above the mark on the hull and/or in a record on the docking plan for the vessel. (The vessel should be provided with a docking plan, a copy of which should be available onboard.)

13 Life-Saving Appliances

13.1 General Requirements

13.1.1 Life-Saving Appliances should be provided in accordance with Table 1 - Life-Saving Appliances.

13.1.2 All equipment fitted should be of a type which has been approved by the Administration in accordance with specification of the particular Schedules of the Merchant Shipping (Life-Saving Appliances) Regulations 1986, SI 1986 No.1066 as amended.

13.1.3 Additional life-saving equipment which is provided should meet the requirements of 13.1.2.

When personal safety equipment is provided for use in water sports activities, arrangements for its stowage should ensure that it will not be used mistakenly as life-saving equipment in an emergency situation.

13.1.4 All life-saving equipment carried should be fitted with retro-reflective material in accordance with the recommendations of Merchant Shipping Notice No. M.1444 - Use and Fitting of Retro-Reflective Material on Life-Saving Appliances - and any Notice amending or replacing it.

13.1.5 Liferaft embarkation arrangements should comply with the following:

.1 Where the distance between the embarkation deck and the top of the liferaft buoyancy tube exceeds 1 metre with the vessel in its lightest condition, an embarkation ladder is to be provided.

.2 Where the distance between the embarkation deck and the top of the liferaft buoyancy tube exceeds 4.5 metres with the vessel in its lightest condition, at least one launching appliance for launching a davit launched liferaft is to be provided on each side of the vessel.

13.1.6 Galvanised steel falls used in launching life-saving appliances should be turned end for end at intervals of not more than 30 months and renewed either when necessary due to deterioration of the falls or at 5 years, whichever is the lesser. However, in lieu of turning "end for end" the Administration may accept a specified period between inspections of the falls and renewal either when necessary due to deterioration or at 4 years, whichever is the lesser.

When falls are of stainless steel, they should be renewed at intervals not exceeding the service life recommended by the manufacturer. Stainless steel falls which do not have a service life recommended by the manufacturer should be treated as galvanised steel falls.

13.1.7 Every inflatable or rigid inflatable rescue boat, inflatable boat, liferaft, inflatable and hydrostatic release unit should be serviced, at intervals not exceeding 12 months, at a service station approved by the manufacturer.

Hydrostatic release units which have been approved for a service life of 2 years and which should be replaced at the end of their life, need not be serviced after 1 year.

13.1.8 Maintenance of equipment should be carried out in accordance with the instructions for onboard maintenance.

13.1.9 The stowage and installation of all life-saving appliances is to be to the satisfaction of the Authorised Surveyor.

13.1.10 All life-saving appliances should be in working order and be ready for immediate use before any voyage is commenced and at all times during the voyage.

13.1.11 For a vessel equipped with stabiliser fins or having other projections at the sides of the hull, special consideration should be given and provisions made as necessary to avoid possible interference with the safe evacuation of the vessel in an emergency.

13.1.12 Means should be provided to prevent overboard discharge of water into survival craft during abandonment.

Table 1 - LIFE-SAVING APPLIANCES (see 13.1)

VESSEL LENGTH IN METRES	≥ 24m < 50m & < 500 GT	≥ 50m OR ≥ 500 GT < 85m	≥ 85m
LIFEBOATS (see 13.2.1)			YES
LIFERAFTS (see 13.2.2)	YES	YES	YES
RESCUE BOAT (see 13.2.3)		YES	YES
INFLATABLE BOAT (see 13.2.3)	YES		
LIFEJACKETS (see 13.2.4)	YES	YES	YES
IMMERSION SUITS (see 13.2.5)	YES	YES	YES/2/0
LIFEBUOYS (TOTAL)	4	8	8
LIFEBUOYS WITH LIGHT AND SMOKE (see 13.2.6.1)	2	2	2
LIFEBUOYS WITH LIGHT		2	2
LIFEBUOYS WITH BUOYANT LINE (see 13.2.6.2)	2	2	2
LINE THROWING APPLIANCE	1	1	1
ROCKET PARACHUTE FLARES	6	12	12
TWO-WAY RADIOTELEPHONE SETS	2	2	3
EPIRB (see 13.2.7)	1	1	1
SART (see 13.2.8)	1	2	2
GENERAL ALARM (see 13.2.9)	YES	YES	YES
LIGHTING (see 13.2.10)	YES	YES	YES
POSTERS AND SIGNS SHOWING SURVIVAL CRAFT AND EQUIPMENT OPERATING INSTRUCTIONS	YES	YES	YES
TRAINING MANUAL	YES	YES	YES
INSTRUCTIONS FOR ONBOARD MAINTENANCE	YES	YES	YES
LIFESAVING SIGNALS AND RESCUE POSTER - SOLAS No 1 IN WHEELHOUSE (see 13.2.11)	YES	YES	YES

13.2 Equipment Carriage Requirements

13.2.1 Lifeboats

13.2.1.1 When lifeboats are required to be carried their acceptance is conditional upon the provision of suitable stowage and launching arrangements.

13.2.1.2 When lifeboats are provided on each side of the vessel, the lifeboat(s) on each side should be of capacity to accommodate the total number of persons onboard.

13.2.1.3 Where it is impractical to carry lifeboats on a vessel, alternative arrangements may be considered as indicated in 13.2.1.4 and 13.2.2.5.

13.2.1.4 For vessels of 85m in length and over, when it is impractical to carry lifeboats on each side of the vessel, alternative arrangements will be considered provided the required subdivision index meets the requirements of 11.3.1.6.

13.2.1.5 A lifeboat will also be acceptable as a rescue boat provided it meets the requirements of Schedule 2 of the Merchant Shipping (Life-Saving Appliances) Regulations 1986.

13.2.2 Liferafts

13.2.2.1 The liferafts carried are to be stowed in GRP containers and must contain the necessary "emergency pack", the contents of which are dependant upon the vessels area of operation (see 13.2.2.8).

13.2.2.2 Liferafts carried should be of equal, or as near equal, capacity as possible.

13.2.2.3 Liferaft approval includes approval of their stowage, launching and float-free arrangements.

13.2.2.4 A vessel should be provided with liferafts of such number and capacity that, in the event of any one liferaft being lost or rendered unserviceable, there is sufficient capacity remaining for all on board.

13.2.2.5 For a vessel of less than 85m in length, one or more liferafts are to be provided on each side of the vessel of sufficient aggregate capacity to accommodate the total number of persons onboard. Liferafts are to be readily transferable for launching on either side of the vessel. (For guidance on the side-to-side transfer of liferafts, reference should be made to Merchant Shipping Notice No. M.1505 - Life-Saving Appliances - The Transfer of Liferafts on Board Ships, and any Notice amending or replacing it.)

If liferafts are not readily transferable, additional liferafts should be fitted so that liferafts having a total capacity of 100% of the vessel's complement are provided on each side of the vessel.

13.2.2.6 For a sailing vessel, when it is impractical to stow the liferafts required by 13.2.2.5 at the ship's side, alternative arrangements can be accepted to provide liferafts having a capacity of 150% of the vessel's complement stowed on the centreline, subject to them being readily transferable to either side of the vessel.

13.2.2.7 When lifeboats are provided in accordance with 13.2.1.2, sufficient liferafts are to be fitted on each side of the vessel capable of accommodating 50% of the total number of persons onboard. Liferafts are to be readily transferable for launching on either side of the vessel (see 13.2.3.5).

If liferafts are not readily transferable, additional liferafts having a total capacity of 100% of the vessel's complement should be provided on each side of the vessel.

13.2.2.8 GRP containers containing liferafts should be stowed on the weatherdeck or in an open space and fitted with hydrostatic release units so that the liferafts will float free of the vessel and automatically inflate.

For a vessel which operates beyond 60 miles from land all the liferafts provided should be equipped with a "SOLAS A PACK". For a vessels which always operate within 60 miles from land, the pack can be a "SOLAS B PACK".

13.2.3 Rescue boat and inflatable boat

13.2.3.1 For a vessel of 50m in length and above or 500 GT and above a rescue boat meeting SOLAS requirements should be provided. The approved rescue boat should have a capacity for not less than 6 persons.

13.2.3.2 The acceptance of an approved rescue boat is conditional upon the provision of suitable stowage and launching arrangements. When a power operated crane is used as a launching device, it should be capable of operation by hand in the event of a power failure.

13.2.3.3 For a vessel of less than 50m in length and less than 500 GT, when an approved rescue boat is not carried onboard alternative arrangements may be considered. These include:

a rescue boat of a SOLAS approved type which is towed by the main vessel; or

a boat which is suitable for rescue purposes carried onboard but which is of a non-SOLAS approved type. The boat should have a capacity for not less than 4 persons and may be a rigid, rigid inflatable or inflatable tender. Tubes of a non-SOLAS inflatable boat should have a minimum of 3 buoyancy compartments built in.

13.2.4 Lifejackets

13.2.4.1 One adult lifejacket should be provided for each person onboard plus spare adult lifejackets sufficient for at least 10% of the total number of persons onboard or two, whichever is the greater. Each lifejacket should be fitted with a light and whistle.

13.2.4.2 Included in the above number of lifejackets there should be at least two inflatable lifejackets for use of the crew of any rescue boat or inflatable boat carried on board.

13.2.4.3 In addition to the adult lifejackets, a sufficient number of children's lifejackets should be provided for children carried on the vessel.

13.2.5 Immersion suits

13.2.5.1 One immersion suit (complying with the requirements of 13.1.2) should be provided for each person onboard.

However, these need not be provided if -

(a) Totally enclosed or partially enclosed lifeboats are fitted; or

(b) Davit launched liferafts are provided; or

(c) The vessel operates all year round on voyages between the parallels of latitude 20° North and 20° South or exempted under 13.2.5.2.

In the case of a vessel which is provided with means for dry-shod emergency evacuation covered by (a) or (b), at least 2 immersion suits should be provided for use by the crew of the rescue boat (see 13.2.2).

13.2.5.2 Owners/managing agents of vessels which operate outside of the parallels of latitude 20° North and 20° South but in areas where the sea water temperature at the time of operation is known and considered to be high enough for dispensation from the safety provision of immersion suits, should apply to the Administration for exemption from the requirements. Full details of the proposed location, period of operation and established temperature data from recognised authorities should be provided.

13.2.6 Lifebuoys

13.2.6.1 Lifebuoys port and starboard provided with combined light and smoke signals should be capable of quick release from the navigating bridge. When this is impractical, they may be stowed at the side of the vessel and provided with conventional release arrangements.

13.2.6.2 The attached buoyant line required on each of two of the lifebuoys is to have a minimum length of 30 metres.

13.2.7 EPIRB

The 406MHz EPIRB should be installed in an easily accessible position ready to be manually released, capable of being placed in a survival craft or floating free if the vessel sinks.

13.2.8 Radar transponders (SART)

The SART is to be stowed in an easily accessible position so that it can be rapidly placed in any survival craft. Means to be provided in order that it can be mounted in the survival craft at a height of at least 1 metre above sea level.

13.2.9 General alarm

13.2.9.1 For a vessel of less than 50m in length and less than 500 GT this alarm may consist of the ship's whistle or siren.

13.2.9.2 For a vessel of 50m in length and above or 500 GT and above the requirement of 13.2.9.1 are to be supplemented by an electrically operated bell or Klaxon system, which is to be powered from the vessel's main supply and also the emergency source of power (see section 8).

13.2.9.3 For a vessel of 85m in length and above, in addition to the requirements of 13.2.9.2, a public address system or other suitable means of communication should be provided.

13.2.10 Lighting

13.2.10.1 Alleyways, internal and external stairways and exits giving access to and including the muster and embarkation stations should be adequately lighted. For a vessel of 50m in length and above or 500 GT and above the lighting should be supplied from the emergency source of power (see section 8).

13.2.10.2 Adequate lighting is to be provided in the vicinity of survival craft, launching appliance(s) (when provided) and the overside area of sea in way of the launching position(s). The lighting should be supplied from the emergency source of power.

13.2.11 Life-Saving signals and rescue poster

When display space in the wheelhouse is restricted, the 2 sides of a SOLAS No.2 poster (as contained in liferaft equipment packs) may be displayed in lieu of a SOLAS No. 1 poster.

14 Fire Safety

14.1 Stowage of Petrol and Other Highly Flammable Liquids

14.1.1 Special consideration should be given to safe conditions of carriage of petrol and other highly flammable liquids either in hand portable containers/tanks or in the tanks of vehicles (such as personal water craft, motor car and helicopter) which may be transported from time to time.

14.1.2 The quantity of petrol and/or other highly flammable liquids carried should be kept to a minimum.

14.1.3 Containers used for the carriage of flammable liquids should be constructed to a recognised standard appropriate to the contents and each container clearly marked to indicate its contents.

14.1.4 Enclosed spaces designated for the safe carriage of petrol or similar fuel or vehicles with fuel in their tanks should be fitted with:-

.1 a fixed fire detection and fire alarm system complying with the requirements of SOLAS regulations II-2/Part A;

.2 a manually activated deluge water spray system of capacity to cover the total area of deck and container/vehicle support platform(s) (if any) at a rate of 3.5l/m^2 per minute, or

for a space in which the provision of a deluge system would be inappropriate/impractical, alternative provisions should be made to the satisfaction of the Administration. (Consideration should be given to the provision of a water spray from at least one fire hose fitted with a jet/spray nozzle being brought to bear on any part of the fuel stowage from the entrance(s) to the space.);

.3 adequate provision for drainage of water introduced to the space by .2. Drainage should not lead to machinery or other spaces where a source of ignition may exist; and

.4 a ducted mechanical continuous supply of air ventilation, which is isolated from other ventilated spaces, to provide at least 6 air changes per hour (based on the empty space) and for which reduction of the airflow should be signalled by an audible and visual alarm on the navigating bridge and at the "in port" control station(s). The ventilation system should be capable of rapid shut down and effective closure in event of fire.

14.1.5 Electrical equipments should be located well clear of those areas where flammable gases are likely to accumulate within the space and be so constructed as to prevent the escape of sparks (ie IP54 as defined in BSEN 60529;1992 'Specification for Degrees of Protection Provided by Enclosures (IP Code)). Electrical equipments less than IP54 should each be provided with an easily accessible and identified means of double pole isolation outside the space, with a fixed flammable gas detector/detectors fitted in the compartment and comprising alarm features on the navigating bridge and elsewhere in the accommodation in accordance with 14B.3.14.2. Where any of these requirements are not practical, then the electrical arrangements should be installed to a suitably certified standard ie flameproof, intrinsically safe etc.

14.1.6 The location of fuel stowages, quantities of fuel and procedures to be followed in an emergency should be approved and recorded on the fire safety plan and/or safety manual, as appropriate.

14.1.7 Reference should be made to section 24 if there is a requirement to provide for helicopter operations to/from the vessel.

14.2 Fire Control Plan(s)

14.2.1 A fire control (general arrangement) plan(s) should be permanently exhibited for the guidance of the Master and crew of the vessel. The content of the plan(s) should adequately show and describe (in English) the principal fire prevention and protection equipment and materials. As far as practical, symbols used on the plans should comply with a recognised international standard.

For each deck, the plan(s) should show the position of control stations; sections of the vessel which are enclosed respectively by "A" class divisions and "B" class divisions; location of flammable liquid storage (see 14.1.4); particulars of and locations of fire alarms, fire detection systems, sprinkler installations, fixed and portable fire extinguishing appliances; fireman's outfit(s); means of access and emergency escapes for compartments and decks; locations and means of control of systems and openings which should be closed down in a fire emergency.

14.2.2 The plan(s) required by 14.2.1 should be kept up to date. Up-dating alterations should be applied to all copies of the plan(s) without delay. Each plan should include a list of alterations and the date on which each alteration was applied.

14.2.3 A duplicate set of the plan(s) should be permanently stored in a prominently marked weathertight enclosure readily accessible to assist non-vessel fire-fighting personnel who may board the vessel in a fire emergency.

14.2.4 Instructions valid to the maintenance and operation of all the equipment and installations onboard for the fighting and containment of fire should be kept in one document holder, readily available in an accessible location.

14A Structural Fire Protection - Vessels of less than 50 metres in length and under 500 GT

14A.1 The boundaries of a space containing internal combustion propulsion machinery or oil fired boilers on a new vessel should be:

.1 gas tight;

.2 capable of preventing the passage of smoke and flame to the end of the 60 minute standard fire test; and

.3 so insulated where necessary with a suitable non-combustible material, that if the division is exposed to a standard fire test, the average temperature on the unexposed side of the division should not increase by more than 139°C above the initial temperature within a period of 30 minutes.

When boundaries are constructed of materials other than steel or aluminium, calculation methods may be used where appropriate to determine compliance with .2 and .3.

14A.2 Fuel tanks and associated pipes and fittings should be located to reduce to a minimum the risk of fire or explosion. Spaces containing such items should be provided with an adequate and efficient ventilation system.

14A.3 In a vessel provided with a gas extinguishing system within an enclosed machinery space, arrangements should be provided for the closure of all openings to the machinery space which can admit air. Additionally, means should be provided for stopping all ventilation fans supplying the machinery space along with the means to cut off the supply of fuel to the engine and any auxiliaries in the event of a fire within the machinery space. The fuel cut off valves should be located as close to the tank as possible (see 7A.2.2).

The above arrangements should be capable of being operated from outside the machinery space.

14A.4 The arrangement of the hull should be such that all underdeck compartments are provided with a satisfactory means of escape. In the case of the accommodation, two means of escape from every restricted space or group of spaces should be provided. Only in an exceptional case should one means of escape be accepted, and then only if the means of escape provided leads directly to the open air and it can be demonstrated that the provision of a second means of escape would be detrimental to the overall safety of the vessel. No escape route should be obstructed by furniture or fittings.

14A.5 When a vessel is not classed by a Classification Society recognised by the Administration and the hull, bulkheads, and main deck are constructed of materials other than steel, evidence of precautions taken to reduce the passage of flame should be submitted to the Administration for approval.

14A.6 Thermal or acoustic insulation fitted should be of a type which is not readily ignitable and, where fitted within a machinery space which does not contain either internal combustion propulsion machinery or oil fired boilers, the surface of the insulation is to be impervious to oil/oil vapour. Insulation provided within a machinery space, which contains either internal combustion propulsion machinery or oil fired boilers, must be of a non-combustible type and the surface of the insulation is also to be impervious to oil/ oil vapour.

14A.7 Paints, varnishes and other finishes which offer an undue fire hazard, should not be used in the engine room or galley or in other areas of high fire risk. Elsewhere such finishes should be kept to a minimum.

14A.8 Upholstery composites (fabric in association with any backing or padding material) used throughout the vessel should satisfy either the cigarette and butane flame tests of British Standard 5852:Part 1:1979 or an equivalent standard, or incorporate an inter-liner which satisfies the ignition resistance test of Schedule 3 of the Consumer Protection, The Furniture and Furnishings (Fire)(Safety) Regulations, 1988.

14A.8.1 Organic foams used in upholstered furniture and mattresses should be of the combustion modified type.

14A.8.2 Suspended textile materials such as curtains or drapes should have type B performance when tested to British Standard 5867:Part 2:1980, or be of equivalent standard.

14A.9 An open flame gas appliance provided for cooking, heating or any other purpose should comply with the requirements of EC Directive 90/396/EEC or equivalent.

The installation of an open flame gas appliance should comply with the appropriate provisions of Annex 4.

14A.10 A fire detection and fire alarm system should be fitted. It should be provided with a control panel located within the wheelhouse, and with audible alarms provided in locations where they are most likely to be heard. The system should comprise smoke, heat or other suitable detectors fitted in the machinery space and galley as a minimum and, in vessels of 30 metres in length and over, suitable detectors should be fitted in all enclosed spaces except those which afford no substantial fire risk.

In the exceptional case of a space/compartment having only one means of escape (14A.4), the integrity of the escape route should be protected by the installation of smoke detectors which should give early warning of danger by means of audible and visible alarms in the space/compartment.

14A.11 Ventilation trunking emanating from either a machinery space or a galley should not, in general, pass through the accommodation spaces. Where this is unavoidable, the trunking should be constructed to the satisfaction of one of the recognised Classification Societies listed in 4.2.1.1. The trunking within the accommodation should be fitted with:

.1 fire insulation to "A-60" standard to a point at least 5 metres from the machinery space or galley; and

.2 automatic fire dampers located in the deck or bulkhead within the accommodation where the trunking passes from the machinery space or galley into the accommodation.

14B Structural Fire Protection - Vessels of 50 metres length and over or 500 GT and over

14B.1 Definitions

Terms used in this section should have the same meaning as defined in SOLAS, except as follows:

"Low flame spread" means that the surface thus described will adequately restrict the spread of flame, this being determined to the satisfaction of the Administration by an established procedure;

"Main vertical zone" means those sections into which the hull, superstructure and deckhouses are divided by A class divisions, the mean length of which on any deck does not normally exceed 40 metres; and

"Not readily ignitable" means that the surface thus described will not continue to burn for more than 20 seconds after removal of a suitable impinging test flame.

14B.2 Existing Vessels

The requirements for vessels of less than 50 metres in length or under 500 GT, as applied to existing vessels, should be complied with and, in addition:

.1 Evidence of precautions taken to reduce the passage of flame throughout accommodation and service spaces should be submitted to the Administration for approval. Such precautions may include the enclosure of stairways and appropriate protection of escape routes.

.2 A fixed fire detection system and fire alarm system of an approved type and complying with the requirements of SOLAS regulations II-2/Part A should be so installed and arranged to detect the presence of fire in all accommodation and service spaces, except spaces which afford no substantial fire risk.

.3 A vessel at all times when at sea, or in port (except when out of service), should be manned or equipped to ensure that any initial fire alarm is immediately received by a responsible member of the crew.

.4 Furniture in corridors and escape routes should be of a type and quantity not likely to obstruct escape.

14B.3 **New Vessels**

New vessels should comply with the following:

14B.3.1 **Ventilation systems**

14B.3.1.1 Ventilation ducts should be of non-combustible material. Short ducts, however, not generally exceeding 2m in length and with a cross-section not exceeding $0.02m^2$ need not be non-combustible, subject to the following conditions:

.1 they should be of a suitable material having regard to the risk of fire;

.2 they should be used only at the end of the ventilation device; and

.3 they should not be situated less than 600mm, measured along the duct, from an opening in an "A" or "B" class division including continuous "B" class ceilings.

14B.3.1.2 Where ventilation ducts with a free cross-sectional area exceeding $0.02m^2$ pass through class "A" bulkheads or decks, the opening should be lined with a steel sheet sleeve unless the ducts passing through the bulkheads or decks are of steel in the vicinity of passage through the deck or bulkhead and the ducts and sleeves should comply in this part with the following:

.1 Sleeves should have a thickness of at least 3mm and a length of at least 900mm. When passing through bulkheads, this length should be divided preferably into 450mm on each side of the bulkhead. The ducts, or sleeves lining such ducts, should be provided with fire insulation. The insulation should have at least the same fire integrity as the bulkhead or deck through which the duct passes.

.2 Ducts with a free cross-sectional area exceeding $0.075m^2$ should be fitted with fire dampers in addition to the requirements of .1 above. The fire damper should operate automatically but should also be capable of being closed manually from both sides of the bulkhead or deck. The damper should be provided with an indicator which shows whether the damper is open or closed. Fire dampers are not required, however, where ducts pass through spaces surrounded by "A" class divisions, without serving those spaces, provided those ducts have the same fire integrity as the divisions which they pierce.

14B.3.1.3 Ducts provided for the ventilation of a machinery space of category A or a galley, should not pass through accommodation spaces, service spaces or control stations unless they comply with the conditions specified in .1 to .4 or .5 and .6 below:

.1 they are constructed of steel having a thickness of at least 3mm and 5mm for duct widths or diameters of up to and including 300mm and 760mm and over respectively and, in the case of ducts with widths or diameters between 300mm and 760mm thickness should be obtained by interpolation;

.2 they are suitably supported and stiffened;

.3 they are fitted with automatic fire dampers close to the boundaries penetrated; and

.4 they are insulated to "A-60" standard from a machinery space or galley to a point at least 5m beyond each fire damper;

or

.5 they are constructed of steel in accordance with .1 and .2 above; and

.6 they are insulated to "A-60" standard throughout accommodation spaces, service spaces or control stations;

except that penetrations of main zone divisions should also comply with the requirements of 14B.3.1.8.

14B.3.1.4 Ducts provided for ventilation to accommodation spaces, service spaces or control stations should not pass through a machinery space of category A or a galley unless they comply with the conditions specified in .1 to .3 or .4 and .5 below:

.1 where they pass through a machinery space of category A or galley, ducts are constructed of steel in accordance with 14B.3.1.3.1 & .2;

.2 automatic fire dampers are fitted close to the boundaries penetrated; and

.3 the integrity of the machinery space or galley boundaries is maintained at penetrations;

or

.4 where they pass through a machinery space of category A or galley, ducts are constructed of steel in accordance with 14B.3.1.3.1 & .2; and

.5 within a machinery space or galley, ducts are insulated to "A-60" standard;

except that penetrations of main zone divisions should also comply with the requirements of 14B.3.1.8.

14B.3.1.5 Ventilation ducts with a free cross-sectional area exceeding $0.02m^2$ passing through "B" class bulkheads should be lined with steel sheet sleeves of 900mm in length divided preferably into 450mm on each side of the bulkheads, unless the duct is of steel for this length.

14B.3.1.6 For a control station outside machinery spaces, practical measures should be taken to ensure that ventilation, visibility and freedom from smoke are maintained so that, in the event of fire, the machinery and equipment contained in the control station may be supervised and continue to function effectively. Alternative and separate means of air supply should be provided; air inlets of the two sources of supply should be so disposed that the risk of both inlets drawing in smoke simultaneously is minimized. These requirements need not apply to control stations situated on, and opening on to, an open deck, or where local closing arrangements would be equally effective.

14B.3.1.7 Exhaust duct(s) from a galley range should be constructed of "A" class divisions where it passes through accommodation spaces and/or spaces containing combustible materials. An exhaust duct should be fitted with:

.1 a grease trap readily removable for cleaning;

.2 a fire damper located in the lower end of the duct;

.3 arrangements for shutting off the exhaust fans, operable from within the galley; and

.4 fixed means for extinguishing a fire within the duct.

14B.3.1.8 When it is necessary for a ventilation duct to pass through a main vertical zone division, a fail-safe automatic closing fire damper should be fitted adjacent to the division. The damper should also be capable of being manually closed from each side of the division. The operating position should be readily accessible and be marked in red light-reflecting colour. The duct between the division and the damper should be of steel or other equivalent material and, if necessary, insulated to comply with the requirements of SOLAS regulation II-2/18.1.1. The damper should be fitted on at least one side of the division with a visible indicator showing whether the damper is in the open position.

14B.3.1.9 Inlets and outlets of ventilation systems should be capable of being closed from outside the space being ventilated.

14B.3.1.10 Power ventilation of accommodation spaces, service spaces, control stations and machinery spaces should be capable of being stopped from an easily accessible position outside the space being served. This position should not be readily cut off in the event of a fire in the spaces served. The means provided for stopping the power ventilation of a machinery space should be entirely separate from the means provided for stopping ventilation of other spaces.

14B.3.2 Structure

14B.3.2.1 The hull, superstructures, structural bulkheads, decks and deckhouses should be constructed of steel or other equivalent material.

14B.3.2.2 However, in cases where any part of the structure is of aluminium alloy, the following should apply:

.1 Insulation of aluminium alloy components of "A" or "B" class divisions, except structure which, in the opinion of the Administration, is non-load-bearing, should be such that the temperature of the structural core does not rise more than 200°C above the ambient temperature at any time during the applicable fire exposure to the standard fire test.

.2 Special attention should be given to the insulation of aluminium alloy components of columns, stanchions and other structural members required to support lifeboat and liferaft stowage, launching and embarkation areas, and "A" and "B" class divisions to ensure that for members:

(a) supporting lifeboat and liferaft areas and "A" class divisions, the temperature rise limitation specified in .1 above should apply at the end of one hour; and

(b) supporting "B" class divisions, the temperature rise limitation specified in .1 above should apply at the end of half an hour.

14B.3.2.3 Crowns and casings of a machinery space of category A should be of steel construction adequately insulated and openings therein, if any, should be suitably arranged and protected to prevent the spread of fire.

14B.3.2.4 In a vessel of less than 1000 GT, crowns and casings of a machinery space of category A need not be of steel provided they are "A-60" divisions and provision is made for boundary cooling through two fire hoses supplied simultaneously from the emergency fire pump with drainage of cooling water overside through scuppers of suitable capacity.

14B.3.3 Main vertical zones and horizontal zones

14B.3.3.1 Hull superstructure and deckhouses in way of accommodation and service spaces should be subdivided into main vertical zones by "A" class divisions. These divisions should have insulation values in accordance with tables 1 and 2.

14B.3.3.2 As far as practical, the bulkheads forming the boundaries of the main vertical zones should be in line with watertight subdivision bulkheads.

14B.3.3.3 Such bulkheads should extend from deck to deck and to the shell or other boundaries.

14B.3.3.4 When a main vertical zone is subdivided by "A" class divisions for the purpose of providing an appropriate barrier between sprinklered and non-sprinklered spaces, the divisions should be insulated in accordance with the fire insulation and integrity values given in tables 1 and 2.

14B.3.4 Bulkheads within a main vertical zone

14B.3.4.1 All bulkheads within accommodation and service spaces which are not required to be "A" class divisions should be at least "B" class or "C" class divisions as prescribed in the tables 1 and 2.

14B.3.4.2 All such divisions may be faced with combustible materials in accordance with the provisions of 14B.3.11.

14B.3.4.3 All corridor bulkheads where not required to be "A" class should be "B" class divisions which should extend from deck to deck except:

.1 when continuous "B" class ceilings or linings are fitted on both sides of the bulkhead, the portion of the bulkhead behind the continuous ceilings or linings should be of material which, in thickness and composition, is acceptable in the construction of "B" class divisions but which should be required to meet "B" class integrity standards only in so far as is reasonable and practical in the opinion of the Administration;

.2 throughout spaces protected by an automatic sprinkler, fire detection and fire alarm system complying with the provisions of 14B.3.13.1.2, the corridor bulkheads of "B" class materials may terminate at a ceiling in the corridor provided such a ceiling is of material which, in thickness and composition, is acceptable in the construction of "B" class divisions. Notwithstanding the requirements of 14B.3.5, such bulkheads and ceilings should be required to meet "B" class integrity standards only in so far as is reasonable and practical. All doors and frames in such bulkheads should be so constructed and erected to provide substantial fire resistance.

14B.3.4.4 All bulkheads required to be "B" class divisions, except corridor bulkheads, should extend from deck to deck and to the shell or other boundaries unless continuous "B" class ceilings or linings are fitted on both sides of the bulkhead, in which case the bulkhead may terminate at the continuous ceiling or lining.

14B.3.5 Fire integrity of bulkheads and decks

14B.3.5.1 In addition to complying with the specific provisions for fire integrity of bulkheads and decks mentioned elsewhere in this section, the minimum fire integrity of bulkheads and decks should be as prescribed in tables 1 and 2.

14B.3.5.2 The following requirements should govern application of the tables:

.1 Tables 1 and 2 should apply respectively to the bulkheads and decks separating adjacent spaces.

.2 For determining the appropriate fire integrity standards to be applied to divisions between adjacent spaces, such spaces are classified according to their fire risk as shown in categories (1) to (9) below. The title of each category is intended to be typical rather than restrictive. The number in parentheses preceding each category refers to the applicable column or row in the tables.

(1) *Control stations*

Spaces containing emergency sources of power and lighting.
Wheelhouse and chartroom.
Spaces containing the vessel's radio equipment.
Fire-extinguishing rooms, fire control rooms and fire-recording stations.
Control room for propulsion machinery when located outside the machinery space.
Spaces containing centralized fire alarm equipment.

(2) *Corridors and lobbies*

(3) *Accommodation spaces*

Spaces so defined, excluding corridors.

(4) *Stairways*

Interior stairways, lifts and escalators (other than those wholly contained within the machinery space(s)) and enclosures thereto.

In this connection, a stairway which is enclosed only at one level should be regarded as part of the space from which it is not separated by a fire door.

(5) *Service spaces (low risk)*

Lockers and store-rooms not having provisions for the storage of flammable liquids and having areas less than $4m^2$ and drying rooms and laundries.

(6) *Machinery spaces of category A*

Spaces so defined.

(7) *Other machinery spaces*

Spaces so defined, excluding machinery spaces of category A.

(8) *Service spaces (high risk)*

Galleys, pantries containing cooking appliances, paint and lamp rooms, lockers and store-rooms having areas of $4m^2$ or more, spaces for the storage of flammable liquids, and workshops other than those forming part of the machinery spaces.

(9) *Open decks*

Open deck spaces and enclosed promenades having no fire risk. Air spaces (the space outside superstructures and deckhouses).

.3 In determining the applicable fire integrity standard of a boundary between two spaces within a main vertical zone or horizontal zone which is not protected by a sprinkler system complying with the provisions of 14B.3.13.1.2 or between such zones neither of which is so protected, the higher of the two values given in the tables should apply.

.4 In determining the applicable fire integrity standard of a boundary between two spaces within a main vertical zone or horizontal zone which is protected by a sprinkler system complying with the provisions of 14B.3.13.1.2 or between such zones both of which are so protected, the lesser of the two values given in the

tables should apply. Where a sprinklered zone and a non-sprinklered zone meet within accommodation and service spaces, the higher of the two values given in the tables should apply to the division between the zones.

14B.3.5.3 Continuous "B" class ceilings or linings, in association with the relevant decks or bulkheads, may be accepted as contributing, wholly or in part, to the required insulation and integrity of a division.

14B.3.5.4 External boundaries which are required to be of steel or other equivalent material may be pierced for the fitting of windows and side scuttles provided that there is no requirement for such boundaries to have "A" class integrity elsewhere in this section. Similarly, in such boundaries which are not required to have "A" class integrity, doors may be of combustible materials, substantially constructed.

Table 1 - Fire integrity of bulkheads separating adjacent spaces

Spaces	(1)	(2)	(3)	(4)	(5)	(6)	(7)	(8)	(9)
Control stations (1)	$A\text{-}0_c$	A-0	A-60	A-0	A-15	A-60	A-15	A-60	*
Corridors and lobbies (2)		C_e	$B\text{-}0_e$	$A\text{-}0_a$ $B\text{-}0_e$	$B\text{-}0_e$	A-60	A-0	A-15 $A\text{-}0_d$	*
Accommodation spaces (3)			C_e	$A\text{-}0_a$ $B\text{-}0_e$	$B\text{-}0_e$	A-60	A-0	A-15 $A\text{-}0_d$	*
Stairways (4)				$A\text{-}0_a$ $B\text{-}0_e$	$A\text{-}0_a$ $B\text{-}0_e$	A-60	A-0	A-15 $A\text{-}0_d$	* *
Service spaces (low risk) (5)					C_e	A-60	A-0	A-0	*
Machinery spaces of category A (6)						*	A-0	A-60	*
Other machinery spaces (7)							$A\text{-}0_b$	A-0	*
Service spaces (high risk) (8)								$A\text{-}0_b$	*
Open decks (9)									

Table 2 - Fire integrity of decks separating adjacent spaces

Spaces above Spaces below	(1)	(2)	(3)	(4)	(5)	(6)	(7)	(8)	(9)
Control stations (1)	A-0	A-0	A-0	A-0	A-0	A-60	A-0	A-0	*
Corridors and lobbies (2)	A-0	*	*	A-0	*	A-60	A-0	A-0	*
Accommodation spaces (3)	A-60	A-0	*	A-0	*	A-60	A-0	A-0	*
Stairways (4)	A-0	A-0	A-0	*	A-0	A-60	A-0	A-0	*
Service spaces (low risk) (5)	A-15	A-0	A-0	A-0	*	A-60	A-0	A-0	*
Machinery spaces of category A (6)	A-60	A-60	A-60	A-60	A-60	*	A-60$_f$	A-60	*
Other machinery spaces (7)	A-15	A-0	A-0	A-0	A-0	A-0	*	A-0	*
Service spaces (high risk) (8)	A-60	A-30 A-0$_d$	A-30 A-0$_d$	A-30 A-0$_d$	A-0	A-60	A-0	A-0	*
Open decks (9)	*	*	*	*	*	*	*	*	-

Notes: To be applied to both tables 1 and 2, as appropriate.

a For clarification on which applies, see 14B.3.4 and 14B.3.7.

b Where spaces are of the same numerical category and subscript $_b$ appears, a bulkhead or deck of the rating shown in the tables is only required when the adjacent spaces are for a different purpose, e.g in category (9), a galley next to a galley does not require a bulkhead but a galley next to a paint room requires an "A-0" bulkhead.

c Bulkheads separating the wheelhouse and chartroom from each other may be "B-0" rating.

d See 14B.3.5.2.3 and 14B.3.5.2.4.

e For the application of 14B.3.3.1, "B-0" and "C", where appearing in table 1, should be read as "A-0".

f Fire insulation need not be fitted if the machinery space in category (7), in the opinion of the Administration, has little or no fire risk.

* Where an asterisk appears in the tables, the division is required to be of steel or other equivalent material but is not required to be of "A" class standard.

For the application of 14B.3.3.1 an asterisk, where appearing in table 2, except for category (9), should be read as "A-0".

14B.3.6 Means of escape

14B.3.6.1 Stairways and ladders should be arranged to provide ready means of escape to the lifeboat and liferaft embarkation deck from all accommodation and service spaces other than machinery spaces. In particular, the following provisions should be complied with:

.1 Below the lowest open deck two means of escape, at least one of which should be independent of watertight doors, should be provided from each watertight compartment, main vertical zone or similarly restricted group of spaces. Exceptionally one of the means of escape may be dispensed with, due regard being paid to the nature and location of spaces and to the number of persons who might normally be accommodated or employed there.

.2 Above the lowest open deck there should be at least two means of escape from each main vertical zone or similarly restricted group of spaces.

.3 Within each main vertical zone there should be at least one readily accessible enclosed stairway providing continuous fire shelter, where practical at all levels up to the appropriate lifeboat and liferaft embarkation decks or the highest level served by the stairway, whichever level is the highest. The width, number and continuity of the stairways should be satisfactory for the number of persons likely to use them.

.4 Access from the stairway enclosures to the lifeboat and liferaft embarkation areas should avoid high fire risk areas.

.5 Stairways serving only a space and a balcony in that space should not be considered as forming one of the required means of escape.

.6 If a radio room or wheelhouse has no direct access to the open deck, two means of escape should be provided, one of which may be a window of sufficient size or another means.

14B.3.6.2 Two means of escape should be provided from each machinery space. In particular, the following provisions should be complied with:

.1 The two means of escape should consist of either:

(a) two sets of steel ladders as widely separated as possible, leading to doors in the upper part of the space similarly separated and from which access is provided to the appropriate lifeboat and liferaft embarkation decks. One of these ladders should provide continuous fire shelter from the lower part of the space to a safe position outside the space; or

(b) one steel ladder leading to a door in the upper part of the space from which access is provided to the embarkation deck and additionally, in the lower part of the space and in a position well separated from the ladder referred to, a steel door capable of being operated from each side and which provides access to a safe escape route from the lower part of the space to the embarkation deck.

.2 One of the means of escape may be dispensed with, due regard being paid to the width and disposition of the upper part of the space.

.3 Two means of escape should be provided from a machinery control room located within a machinery space, at least one of which should provide continuous fire shelter to a safe position outside the machinery space.

14B.3.6.3 In no case should lifts be considered as forming one of the required means of escape.

14B.3.7 Protection of stairways and lifts in accommodation and service spaces

14B.3.7.1 A stairway should be of steel frame construction except where the Administration sanctions the use of other equivalent material, and should be within enclosures formed of "A" class divisions, with positive means of closure at all openings, except that:

.1 a stairway which penetrates a single deck only may be protected at one level only by at least "B" class divisions and self-closing door(s); and

.2 stairways may be fitted in the open in a public space, provided they lie wholly within such public space.

14B.3.7.2 A stairway enclosure should have direct communication with the corridors and be of sufficient area to prevent congestion, having in view the number of persons likely to use them in an emergency. In so far as is practical, stairway enclosures should not give direct access to cabins, service lockers, or other enclosed spaces containing combustibles in which a fire is likely to originate.

14B.3.7.3 A lift trunk should be so fitted to prevent the passage of flame from one 'tween-deck to another and should be provided with means of closing to permit the control of draught and smoke.

14B.3.8 Openings in "A" class divisions

14B.3.8.1 Except for hatches between store and baggage spaces, and between such spaces and the weather decks, all openings should be provided with permanently attached means of closing which should be at least as effective for resisting fires as the divisions in which they are fitted.

14B.3.8.2 The construction of all doors and door frames in "A" class divisions, with the means of securing them when closed, should provide resistance to fire as well as the passage of smoke and flame, as far as practical, equivalent to that of the bulkheads in which the doors are situated. Such doors and door frames should be constructed of steel or other equivalent material. Watertight doors need not be insulated.

14B.3.8.3 It should be possible for each door to be opened and closed from each side of the bulkhead by one person only.

14B.3.8.4 Fire doors in main vertical zone bulkheads and stairway enclosures, other than power-operated watertight doors and those which are normally locked, should be of the self-closing type capable of closing against an inclination of 3.5° opposing closure. The speed of door closure should, if necessary, be controlled to prevent undue danger to persons. All such doors, except those that are normally closed, should all be capable of release from a control station, either simultaneously or in groups, and also individually from a position at the door. The release mechanism should be so designed that the door will automatically close in the event of disruption of the control system; however, approved power-operated watertight doors will be considered acceptable for this purpose. Hold-back hooks not subject to control station release should not be permitted. When double swing doors are permitted, they should have a latch arrangement which is automatically engaged by the operation of the door release system.

14B.3.8.5 When a space is protected by an automatic sprinkler system complying with the provisions of 14B.3.13.1.2 or fitted with a continuous "B" class ceiling, openings in decks not forming steps in main vertical zones nor bounding horizontal zones should be closed reasonably tight and such decks should meet the "A" class integrity requirements in so far as is reasonable and practical.

14B.3.9 Openings in "B" class divisions

14B.3.9.1 Doors and door frames in "B" class divisions and means of securing them should provide a method of closure which should have resistance to fire as far as practical equivalent to that of the divisions except that a ventilation opening may be permitted in the lower portion of such doors. When such an opening is in or under a door the total net area of the opening(s) should not exceed $0.05m^2$. When such an opening is cut in a door it should be fitted with a grill made of non-combustible material. Doors should be non-combustible or of substantial construction.

14B.3.9.2 When a sprinkler system complying with the provisions of 14B.3.13.1.2 is fitted:

.1 openings in decks not forming steps in main vertical zones nor bounding horizontal zones should be closed reasonably tight and such decks should meet the "B" class integrity requirements in so far as is reasonable and practical in the opinion of the Administration; and

.2 openings in corridor bulkheads of "B" class materials should be protected in accordance with the provisions of 14B.3.4.

14B.3.10 Windows and side scuttles (Also see 5.4 and 5.5)

14B.3.10.1 All windows and side scuttles in bulkheads within accommodation, service spaces and control stations should be so constructed to preserve the integrity requirements of the type of bulkheads in which they are fitted.

14B.3.10.2 Notwithstanding the requirements of tables 1 and 2 all windows and side scuttles in bulkheads separating accommodation and service spaces and control stations from weather should be constructed with frames of steel or other suitable material. The glass should be retained by a metal glazing bead or angle.

14B.3.11 Restricted use of combustible materials

14B.3.11.1 Except in baggage rooms, or refrigerated compartments of service spaces, all linings, grounds, draught stops, ceilings and insulations should be of non-combustible materials.

14B.3.11.2 Vapour barriers and adhesives used in conjunction with insulation, as well as insulation of pipe fittings, for cold service systems need not be non-combustible, but they should be kept to the minimum quantity practicable and their exposed surfaces should be low flame spread.

14B.3.11.3 The following surfaces should be low flame spread:

.1 exposed surfaces in corridors and stairway enclosures, and of bulkheads, wall and ceiling linings in all service spaces and control stations;

.2 concealed or inaccessible spaces in accommodation, service spaces and control stations; and

.3 exposed surfaces of bulkheads, wall and ceiling linings in accommodation spaces not protected by a system complying with 14B.3.13.1.2 or 14B.3.13.1.3.

14B.3.11.4 The total volume of combustible facings, mouldings, decorations and veneers in any accommodation and service space not protected by a system complying with 14B.3.13.1.2 or 14B.3.13.1.3, should not exceed a volume equivalent to 2.5mm veneer on the combined area of the walls and ceilings.

14B.3.11.5 Veneers used on surfaces and linings covered by the requirements of 14B.3.11.3 should have a calorific value not exceeding $45MJ/m^2$ of the area for the thickness used.

14B.3.11.6 Furniture in the corridors and escape routes should be of a type and quantity not likely to obstruct access.

14B.3.11.7 Primary deck coverings, if applied within accommodation and service spaces and control stations, should be of material which will not readily ignite.

14B.3.11.8 Upholstery composites (fabric in association with any backing or padding material) used throughout the vessel should satisfy either the cigarette and butane flame test of British Standard 5852:Part 1:1979 or an equivalent standard, or incorporate an inter-liner which satisfies the ignition resistance test of Schedule 3 of the Consumer Protection, The Furniture and Furnishings (Fire)(Safety) Regulations 1988; and

.1 Organic foams used in upholstered furniture and mattresses should be of the combustion modified type; and

.2 Suspended textile materials such as curtains and drapes should have a type B performance when tested to British Standard 5867:Part 2:1980 or be of equivalent standard.

14B.3.12 Details of construction

14B.3.12.1 In accommodation and service spaces, control stations, corridors and stairways:

.1 air spaces enclosed behind ceilings, panelling or linings should be suitably divided by close-fitting draught stops not more than 7m apart; and

.2 in the vertical direction, enclosed air spaces, including those behind linings of stairways, trunks, etc. should be closed at each deck.

14B.3.12.2 Without impairing the efficiency of the fire protection, the construction of ceilings and bulkheads should allow a fire patrol to detect any smoke originating in concealed and inaccessible places, except where there is no risk of fire originating in such places.

14B.3.12.3 When gaseous fuel is used for domestic purposes, the arrangements for the storage, distribution and utilisation of the fuel should be such that, having regard to the hazards of fire and explosion which the use of such fuel may entail, the safety of the vessel and the persons onboard are preserved.

In particular, open flame gas appliances provided for cooking, heating or any other purposes, should comply with the requirements of EC Directive 90/396/EEC or equivalent and, the installation of open flame gas appliances should comply with the appropriate provisions of Annex 4.

14B.3.13 Fixed fire detection and fire alarm systems and automatic sprinkler, fire detection and fire alarm systems

14B.3.13.1 Each separate zone in all accommodation and service spaces, except spaces which afford no substantial fire risk such as void spaces, sanitary spaces, etc., and, having regard for 14B.3.11.4, should be provided throughout with either:

.1 a fixed fire detection and fire alarm system of an approved type and complying with the requirements of SOLAS regulation II-2/13, installed and arranged to detect the presence of fire in such spaces; or

.2 an automatic sprinkler, fire detection and fire alarm system of an approved type and complying with the requirements of SOLAS regulation II-2/12 and installed and arranged to protect such spaces. In addition, a fixed fire detection and fire alarm system of an approved type complying with the requirements of SOLAS regulation II-2/13 should be installed and arranged to provide smoke detection in corridors, stairways and escape routes within accommodation spaces; or

.3 a manual dry pipe sprinkler system of an approved type either complying with the requirements of IMO Resolution MSC.44(65) or, to the satisfaction of the Administration, an equivalent standard which provides increased security against damage caused by accidental discharge from sprinklers. The system should be installed and arranged to protect such spaces. In addition, a fixed addressable fire detection and alarm system of an approved type complying with the requirements of SOLAS regulation II-2/13 should be installed and arranged to detect the presence of fire in such spaces.

14B.3.14 Fire detection and alarms

14B.3.14.1 Manually operated call points complying with the requirements of SOLAS regulation II-2/13 should be installed.

14B.3.14.2 At all times, vessels when at sea and in port (except when out of service) should be manned and/or equipped to ensure that any initial fire alarm is immediately received by a responsible member of the crew.

15A Fire Appliances - Vessels of less than 50 metres in length and under 500 GT

15A.1 General Requirements

15A.1.1 Fire appliances of an approved type should be provided at least to the extent listed in Table 1 and, in any case, to the satisfaction of the Administration and the specific requirements of 15A.2.

15A.1.2 Fire appliances provided in addition to those required by 15.1.1 should be of an approved type.

15A.1.3 The location, installation, testing and maintenance of all equipment should be to the satisfaction of the Administration.

15A.2 Specific Requirements

15A.2.1 Provision of water jet

At least one jet of water should be able to reach any part of the vessel normally accessible to passengers or crew while the vessel is being navigated and, any store room and any part of a storage compartment when empty.

15A.2.2 Fire pumps

15A.2.2.1 The power driven fire pump should have a capacity of -

$$2.5 \times \{1+0.066 \times (L(B+D))^{0.5}\}^2 \text{ m}^3\text{/hour}$$

where: L is the length;
B is the greatest moulded breadth; and
D is the moulded depth measured to the bulkhead deck at amidships.

When discharging at full capacity through 2 adjacent fire hydrants, the pump should be capable of maintaining a water pressure of 0.2N/mm^2 at any hydrant, provided the fire hose can be effectively controlled at this pressure.

Table 1 - FIRE APPLIANCES - VESSELS OF LESS THAN 50 METRES IN LENGTH AND UNDER 500 GT (see 15A.1.1)

1	PROVISION OF WATER JET - sufficient to reach any part of vessel	1
2	POWER DRIVEN FIRE PUMP - engine or independent drive	1
3	ADDITIONAL HAND OR INDEPENDENT POWER DRIVEN FIRE PUMP AND ITS SEA CONNECTION - not located in the same space as item 2 or a machinery space containing internal combustion type machinery	1
4	FIRE MAIN & HYDRANTS - to achieve item 1 with a single length of hose	Sufficient
5	HOSES - with jet/spray nozzles each fitted with a shut-off facility	3
6	FIRE EXTINGUISHERS - portable, for use in accommodation and service spaces	3
7	FIRE EXTINGUISHERS - for a machinery space containing internal combustion type machinery - the options are: (a) a fixed fire extinguishing system complying with SOLAS regulations II-2/Part A; and	
	(b) (i) 1 portable extinguisher for oil fires for each 74.6kW power; or	7 (max)
	(ii) 2 portable extinguishers for oil fires together with either - 1 foam extinguisher of 45l capacity; or 1 CO_2 extinguisher of 16kg capacity	2 + 1
8	FIREMANS OUTFIT - to include one breathing apparatus of the air hose type	1
9	FIRE BLANKET - in galley	1

15A.2.2.2 The second fire pump should have a capacity -

for a hand pump, sufficient to produce a throw of at least 6 metres through a fire hose with a 10mm diameter nozzle and which can be directed on any part of the vessel; or

for a power driven pump, at least 80% of that required by 15A.2.2.1 and be fitted with an input to the fire main.

15A.2.3 Fire main and hydrants

15A.2.3.1 A fire main, water service pipes and fire hydrants should be fitted.

15A.2.3.2 The fire main and water service pipe connections to the hydrants should be sized for the maximum discharge rate of the pump(s) connected to the main.

15A.2.3.3 The fire main, water service pipes and fire hydrants should be constructed such that they will:

.1 not be rendered ineffective by heat;

.2 not corrode; and

.3 be protected against freezing.

15A.2.3.4 When a fire main is supplied by 2 pumps, 1 in the machinery space and 1 elsewhere, provision should be made for isolation of the fire main within the machinery space and for the second pump to supply the fire main and hydrants external to the machinery space.

The isolation valve(s) should be fitted outside the machinery space in a position easily accessible in the event of a fire.

15A.2.3.5 The fire main should have no connections other than those necessary for fire fighting or washing down.

15A.2.3.6 Fire hydrants should be located for easy attachment of fire hoses, protected from damage and distributed so that the fire hoses provided can reach any part of the vessel.

15A.2.3.7 Fire hydrants should be fitted with valves which allow a fire hose to be isolated and removed when a fire pump is working.

15A.2.4 Fire hoses

15A.2.4.1 Fire hoses should not exceed 18 metres in length and, generally, the diameter of a lined hose for use with a powered pump should not be less than 45mm.

15A.2.4.2 Fire hoses and associated tools and fittings should be kept in readily accessible and known locations close to the hydrants or connections on which they will used. Hoses supplied from a powered pump should have jet/spray nozzles (incorporating a shut-off facility) of diameter 19mm, 16mm or 12mm depending on fire fighting purposes. (For accommodation and service spaces, the diameter of nozzles need not exceed 12mm.)

15A.2.4.3 Hydrants or connections in interior locations on the vessel should have hoses connected at all times. For use within accommodation and service spaces proposals to provide smaller diameter of hoses and jet/spray nozzles will be considered.

15A.2.4.4 The number of fire hoses and nozzles provided should correspond to the functional fire safety requirements but, be at least 3.

15A.2.5 Portable fire extinguishers for use in accommodation and service spaces

15A.2.5.1 The number, location, fire extinguishing medium type and capacity should be selected according to the perceived fire risk but, at least 3 portable fire extinguishers should be provided. As far as practical, the fire extinguishers provided should have a uniform method of operation.

15A.2.5.2 Portable fire extinguishers of carbon dioxide type should not be located or provided for use in accommodation spaces.

15A.2.5.3 Except for portable extinguishers provided in connection with a specific hazard within a space when it is manned (such as a galley), portable extinguishers generally should be located external to but adjacent to the entrance of the space(s) in which they will be used. Extinguishers should be stowed in readily accessible and marked locations.

15A.2.5.4 Spare charges should be provided onboard for at least 50% of each type and capacity of portable fire extinguisher onboard. When an extinguisher is not of a type which is rechargeable when the vessel is at sea, an additional portable fire extinguisher of the same type (or its equivalent) should be provided.

15A.2.6 Fire extinguishing in machinery spaces

15A.2.6.1 In a machinery space containing internal combustion type machinery fire appliances should be provided at least to the extent listed in item 7 of Table 1 - Fire Appliances.

15A.2.6.2 In a machinery space containing an oil fired boiler, oil fuel settling tank or oil fuel unit, a fixed fire extinguishing system complying with SOLAS regulations II-2/Part A should be installed.

15A.2.6.3 Portable fire extinguishers should be installed and the number, location, fire extinguishing medium type and capacity should be selected according to the perceived fire risk in the space. (Spare charges or spare extinguishers should be provided per 15A.2.5.4.)

In any case, portable fire extinguishers for extinguishing oil fires should be fitted:

.1 in a boiler room - at least 2;

.2 in a space containing any part of an oil fuel installation - at least 2; and

.3 in a firing space - at least 1.

15B Fire Appliances - Vessels of 50 metres in length and over or 500 GT and over

15B.1 Existing Vessels

An existing vessel should comply with the requirements for a new vessel so far as it is reasonable and practical to do so but, in any case, at least the requirements for a vessel of less than 50 metres in length and under 500 GT.

15B.2 New Vessels

A new vessel should comply with SOLAS regulations II-2/Part A as appropriate to the vessel and its equipment. For the purpose of the SOLAS regulations the standards for a cargo ship apply.

In no case should the standards applied be less than those applied to a vessel of less than 50 metres in length and under 500 GT.

16 Radio

16.1 General

16.1.1 All vessels regardless of size should comply with the requirements of this chapter.

16.2 Radiocommunications: The Global Marine Distress and Safety System (GMDSS)

16.2.1 Each vessel should carry sufficient radio equipment to perform the following distress and safety communications functions throughout its intended voyage:

.1 transmitting ship to shore distress alerts by at least two separate and independent means, each using a different radiocommunication service;

.2 receiving shore-to-ship distress alerts;

.3 transmitting and receiving ship-to-ship distress alerts;

.4 transmitting and receiving search and rescue co-ordinating communications;

.5 transmitting and receiving on-scene communications;

.6 transmitting and receiving signals for locating by radar;

.7 transmitting and receiving maritime safety information; and

.8 transmitting and receiving bridge-to-bridge communications.

16.2.2 Radio installations

16.2.2.1 Table 1 illustrates the radio installations to be carried to fulfil the functional requirements sailing at different distances from a safe haven.

Table 1

Radio Equipment	Distance from safe haven - nautical miles			
	Up to 30	Up to 60	Up to 150	Unlimited
VHF Radiotelephone with Digital Selective Calling (DSC)	One	One	One	One
MF/HF Radiotelephone with Digital Selective Calling (DSC)	None	None	One[1]	One[1]
INMARSAT Ship Earth Station	None	None	One[1]	One[1]
NAVTEX[2] receiver	None	None	One	One

Notes:

1. An INMARSAT ship earth station OR an MF/HF radiotelephone with DSC may be fitted for operations over 60 miles from a safe haven.

2. If the vessel is sailing in an area where an international NAVTEX service is not provided then the NAVTEX receiver should be substituted by an INMARSAT enhanced group calling system.

16.2.2.2 *Example*

As an illustration, the minimum equipment to be installed on a vessel engaged on world-wide operations south of 70° North and north of 70° South would be:

one VHF radiotelephone with DSC;
one INMARSAT-C ship earth station;
one NAVTEX receiver.

Note also the requirement for the carriage of two-way radiotelephone sets, EPIRB's and SART's given in section 13 table 1.

16.2.3 Operational performance

16.2.3.1 All radiocommunications equipment should be of a type which is approved by the relevant authority.

16.2.4 Installation

16.2.4.1 The radio installation should:

.1 be so located to ensure the greatest possible degree of safety and operational availability;

.2 be protected against harmful effect of water, extremes of temperature and other adverse environmental conditions;

.3 be clearly marked with the call sign, the vessel station identity and any other codes applicable to the use of the radio installation.

16.2.5 Sources of energy

16.2.5.1 There should be available at all times, while the vessel is at sea, a supply of electrical energy sufficient to operate the radio installations and to charge any batteries used as part of a reserve source or sources of energy for the radio installations.

16.2.5.2 A reserve source or energy, independent of the propelling power of the vessel and its electrical system, should be provided for the purpose of conducting distress and safety radiocommunications for a minimum of one hour in the event of failure of the vessel's main and, if provided, emergency sources of electrical power.

16.2.5.3 When a reserve source of energy consists of a rechargeable accumulator battery, a means of automatically charging such batteries should be provided which is capable of recharging them to minimum capacity requirements within 10 hours.

16.2.5.4 The siting and installation of accumulator batteries should ensure the highest degree of service and safety.

16.2.6 Watches

16.2.6.1 A vessel, while at sea, should maintain a continuous watch:

.1 where practicable, on VHF Channel 16;

.2 where practicable, on VHF Channel 13;

.3 on VHF Digital Selective Calling (DSC), on channel 70;

.4 if fitted with an MF radiotelephone, on 2182kHz and, on the distress and safety DSC frequency 2187.5kHz;

.5 for satellite shore-to-ship distress alerts if fitted with a radio facility for reception of maritime safety information by the INMARSAT enhanced group calling system;

.6 for broadcasts of maritime safety information on the appropriate frequency or frequencies on which such information is broadcast for the area in which the vessel is navigating; normally using the International NAVTEX service or INMARSAT's enhanced group calling facility. (Further information may be obtained from the Admiralty List of Radio Signals volume 3.)

16.2.7 **Radio personnel**

16.2.7.1 A vessel should carry at least one person qualified for distress and safety radiocommunication purposes, who should hold a certificate of competence acceptable to the relevant authority.

17 Navigation Lights, Shapes and Sound Signals

Every vessel should comply with the requirements of the International Regulations For Preventing Collisions At Sea, 1972, as amended.

18 Navigational Equipment and Visibility from Wheelhouse

18.1 Navigational Equipment

18.1.1 A vessel should be fitted with an efficient magnetic compass complying with the following requirements, as appropriate:

.1 On a steel vessel it should be possible to correct the compass for co-efficients B, C and D.

.2 The magnetic compass or a repeater should be so positioned as to be clearly readable by the helmsman at the main steering position. It should also be provided with an electric light, the electric power supply to be twin wire type.

.3 Means should be provided for taking bearings as near as practicable over an arc of the horizon of 360°. This requirement may be met by the fitting of a pelorus or, on a vessel other than a steel vessel, a hand bearing compass.

18.1.2 A vessel should be fitted with an echo sounder.

18.1.3 A vessel should be provided with, the following additional equipment:

.1 an electronic navigational positioning system appropriate to the area of operation;

.2 a distance measuring log;

.3 a gyro compass or spare magnetic compass bowl;

.4 a rudder angle indicator; and

.5 9 GHz radar.

18.2 Look-out

18.2.1 The normal conning position should permit a good all round view of the horizon. The sea surface should not be obscured for more than two vessel lengths forward of the bow.

18.2.2 Windows to the navigating position should not be of either polarised or tinted glass (see 5.5.5). Portable tinted screens may be provided for selected windows.

19 Miscellaneous Equipment

19.1 Nautical Publications

Every vessel should comply with the requirements of the Merchant Shipping (Carriage of Nautical Publications) Rules 1975, SI 1975 No.700, as amended.

19.2 Signalling Lamp

Every vessel should carry an approved signalling lamp.

19.3 Measuring Instruments

Every vessel should carry a barometer.
Every sailing vessel should carry an anemometer and an inclinometer.

19.4 Searchlight

Every vessel should carry an efficient fixed or portable searchlight suitable for manoverboard search and rescue operations. This may be the approved signalling lamp required by 19.2.

20 Anchors and Cables and Towing Arrangements

20.1 Equipment

20.1.1 Vessels will be considered to have adequate equipment if fitted out in accordance with certification standards set by any of the Classification Societies listed in 4.2.1.1.

20.1.2 Vessels not built in accordance with 20.1.1 may be specially considered by the Administration, provided full information is submitted for approval.

20.2 Sailing Vessels

20.2.1 The sizing of anchors and cables for sailing vessels should take into account the additional windage effect of the masts and rigging.

20.2.2 Typically, for square rigged sailing vessels, experience based guidance on approximate increase in anchor mass and cable strength required is:

for vessels up to 50 metres in length, typically 50% above the requirements for a typical motor vessel having the same total longitudinal profile area of hull and superstructure as the square rigged sailing vessel under consideration; and

for vessels 100 metres in length and over, typically 30% above the requirements for a typical motor vessel having the same total longitudinal profile area of hull and superstructure as the square rigged sailing vessel under consideration.

For a square rigged sailing vessel of between 50 and 100 metres in length the increase should be obtained by linear interpolation.

20.3 Towing Arrangements

Accessible efficient strong securing points should be provided for the attachment of tow lines for the vessel to tow and be towed.

21 Accommodation

21.1 General

An adequate standard of accommodation should be provided to ensure the comfort, recreation, health and safety of all persons on board.

Attention is drawn to the achievement of appropriate standards for means of access and escape, lighting, heating, food preparation and storage, safety of movement about the vessel, ventilation and water services.

Generally, accommodation standards for the crew should be at least equivalent to the standards set by the International Labour Organisation conventions for crew accommodation in merchant ships. When it is neither reasonable nor practicable to site crew sleeping accommodation amidships or aft and above the deepest waterline as required, measures taken to ensure an equivalent level of crew health and safety should be agreed with the Administration.

The following standards are described by general principles which need to be expanded to meet the requirements which relate to the use and areas of operation of particular vessels.

21.2 Access/Escape Arrangements

See 14A.5 and 14B.3.6.

21.3 Lighting

An electric lighting system should be installed which is capable of supplying adequate light to all enclosed accommodation and working spaces. The system should be designed and installed in accordance with section 8.

21.4 Heating

As considered appropriate, an adequate heating installation should be provided.

21.5 Food Preparation and Storage

The galley should be provided with a cooking stove fitted with fiddle bars and a sink, and have adequate working surface for the preparation of food.

The galley floor should be provided with a non-slip surface and provide a good foothold.

When a cooking appliance is gimballed, it should be protected by a crash bar or other means to prevent personal injury. Means should be provided to lock the gimballing mechanism.

Means should be provided to allow the cook to be secured in position, with both hands free for working, when the vessel motions threaten safe working.

Secure and hygienic storage for food should be provided.

21.6 Hand Holds and Grab Rails

There should be sufficient hand holds and grab rails within the accommodation to allow safe movement around the accommodation at all times.

21.7 Ventilation

Effective means of ventilation should be provided to all enclosed spaces which are entered by personnel.

Mechanical ventilation should be provided to all accommodation spaces on vessels which are intended to make long international voyages or operate in tropical waters.

As a minimum, mechanical ventilation should be capable of providing 6 changes of air per hour, when all access and other openings (other than ventilation intakes) to the spaces are closed.

21.8 Water Services

An adequate supply of fresh drinking water should be provided and piped to convenient positions throughout the accommodation spaces.

In addition, an emergency reserve supply of drinking water should be carried, sufficient to provide at least 2 litres per person.

21.9 Sleeping Accommodation

A bed (bunk or cot) should be provided for every person on board. Where considered appropriate, means for preventing the occupants from falling out, should be provided.

21.10 Toilet Facilities

Adequate sanitary toilet facilities should be provided on board. The facilities should be at least one water closet, one wash hand basin and one shower for every 12 persons or part thereof.

In vessels where a sanitary system, including a holding tank, are provided, care should be taken to ensure that there is no possibility of fumes from the tank finding their way back to a toilet, should the water seal at the toilet be broken.

21.11 Stowage Facilities for Personal Effects

Adequate stowage facilities for clothing and personal effects should be provided for every person on board.

21.12 Securing of Heavy Equipment

All heavy items of equipment such as ballast, batteries, cooking stove, etc, should be securely fastened in place. All stowage lockers containing heavy items should have lids or doors which are capable of being securely fastened.

22 Protection of Personnel

22.1 Deckhouses and Superstructures

The structural strength of any deckhouse or superstructure should comply with the requirements of one of the Classification Societies listed in 4.2.1.1, as appropriate to the vessel and its areas of operation.

22.2 Bulwarks and Guard Rails

22.2.1 Wherever reasonable and practicable, vessels should comply with ICLL requirements and the interpretation used by any of the Classification Societies listed in 4.2.1.1.

22.2.2 When it can be justified that such standards cannot be complied with, compliance with the following guidelines may be considered by the Administration:

.1 Where the function of the vessel is not impeded and there are frequently people on the deck, bulwarks or three courses of rail or taut wires should be fitted around the deck at a height of not less than 1000mm above the deck. They should be supported at intervals not exceeding 2.2 metres. Intermediate courses of rails or wires should be evenly spaced.

.2 Where the function of the vessel would be impeded by the provision of bulwarks and/or guard rails complying with 22.2.1, alternative proposals detailed to provide equivalent safety for persons on deck should be submitted to the Administration for approval.

22.3 Safe Work Aloft and on the Bowsprit of Sailing Vessels

22.3.1 When access to the rig is an operational necessity, provision should be made to enable people to work safely aloft and out on the bowsprit, to the satisfaction of the Administration.

22.3.2 The arrangements provided should be based on established safe working practices for the type of vessel. The arrangements may include but not be limited to:-

.1 Safety nets below the bowsprit.

.2 Safety grabrails in wood (or jackstays in metal) fixed along the bowsprit to act as handholds and safety points for safety harnesses.

.3 Mandatory use of safety harnesses aloft and for work on the bowsprit.

.4 Sufficient footropes and horses in wire (or rope) permanently rigged to enable seamen to stand on them whilst working out on the yards or on the bowsprit.

.5 Safety jackstays (in metal) fixed along the top of the yards, to provide handholds and act as strong points for safety harnesses.

.6 Means of safely climbing aloft, such as:-

(i) fixed metal steps or ladders attached to the mast; or

(ii) traditional ratlines (rope) or, rattling bars (wood/steel), fixed across the shrouds to form a permanent ladder.

22.3.3 A vessel should be provided with a "Training Manual" which should include details of established safe working practices specific to the vessel, guidance on training for members of the crew and personal clothing and protection from injury.

22.4 Recovery of Persons from the Sea

22.4.1 Means should be provided for the recovery of a person from the sea to the vessel. The means should allow that the person is unconscious or unable to assist in the rescue.

The means of recovery should be demonstrated to the satisfaction of an Authorised Surveyor.

22.4.2 If an overside boarding ladder or scrambling net is provided to assist in the recovery of an unconscious person from the water, the ladder or net should extend from the weather deck to at least 600mm below the lowest operational waterline.

22.5 Personal Clothing

It should be the responsibility of the owner/managing agent to ensure that the following requirements and recommendations for items of personal clothing are discharged:

.1 All persons on board should have suitable protective clothing and equipment appropriate to the prevailing air and sea temperatures and weather conditions.

.2 It is strongly recommended that all persons on board wear footwear provided with non-slip soles, particularly on the open deck.

22.6 Noise

22.6.1 Attention is drawn to the second edition of the "Code of practice for noise levels in ships" (Noise Code) published by HMSO in 1990. (The Noise Code contains the IMO Code on Noise Levels on Board Ships - Resolution A.468(XII).)

22.6.2 New vessels covered by this Code should meet the recommendations of the Noise Code so far as reasonable and practicable.

22.6.3 Existing vessels should be considered with particular regard to the recommendations of the Noise Code for protection of the crew from noise levels which may give rise to noise-induced hearing loss.

22.6.4 The Noise Code recognises that the scope for strict application of recommended noise levels on small vessels can be limited and deals with the means of protecting the seafarer from the risk of noise-induced hearing loss under conditions where, at the present time, it is not technically feasible to limit the noise to a level which is not potentially harmful.

22.6.5 For safe navigation, so that sound signals and VHF communications can be heard, it is recommended that a noise level of 65dB(A) at the navigating position is not exceeded.

22.6.6 For machinery spaces, workshops and stores which are manned either continuously or for lengthy periods, the recommended limits are 90dB(A) (see 22.6.7) for machinery spaces and 85dB(A) for workshops and stores. For any space which is required to be manned and in which the noise level exceeds 90dB(A), consideration should be given to the provision of a designated refuge from noise.

For machinery spaces which are not intended to be continuously manned or are attended for short periods only, the recommended limit is 110dB(A) (see 22.6.7).

The limits have been set from hearing damage risk considerations and the use of suitable ear protectors.

22.6.7 To indicate the need to wear ear protectors in spaces in which the noise level exceeds 85dB(A), each entrance to the space should be provided with a warning notice comprising a symbol complying with British Standards Institution specification BS 5378:1980 and supplementary sign stating "High Noise Levels. Use Ear Protectors". Efficient ear protectors should be provided for use in such spaces.

23 Medical Stores

A vessel should carry medical stores as required by the Merchant Shipping and Fishing Vessels (Medical Stores) Regulations 1995, SI 1995 No.1802 as amended. (The Regulations refer to Merchant Shipping Notice No. M.1607 for details of medicines and medical stores to be carried.)

Medical training requirements for members of the complement of the vessel are given in section 26.

24 Shore-ship Transfer of Personnel

24.1 Tenders (Dinghies)

24.1.1 When a vessel carries a rigid or inflatable tender in addition to any lifeboat/rescue boat/ inflatable boat which may be carried in compliance with life-saving appliance requirements, the tender should be fit for its intended use.

24.1.2 Safety equipment should be provided in the tender as appropriate to its intended range and areas of operation.

24.1.3 Each tender should be clearly marked with the number of persons (mass 75 kg) that it can safely carry, and the name of the parent vessel.

24.1.4 Inflatable tenders should be of a type which have a minimum of 3 buoyancy compartments built into them, and should be maintained in a safe condition by the operator.

24.1.5 In the case of petrol engined tenders, see section 14 for the safety requirements for the carriage of petrol.

24.2 Helicopter

24.2.1 When provision is made for helicopter operations to/from the vessel, the arrangements should comply with the requirements of SOLAS regulation II-2/18 which apply to helicopter facilities and the standards set for landing and other operational matters by the International Chamber of Shipping publication "Guide to Helicopter/Ship Operations".

N.B. Proposed amendments to the SOLAS requirements which apply to helicopter facilities are awaiting finalisation and adoption by the IMO.

24.2.2 If it is proposed to provide refuelling facilities for a helicopter whilst it is onboard the vessel, consideration and approval of the proposal should be considered by the Administration on an individual ship basis.

24.3 Pilot for Vessel

Boarding arrangements provided for pilots should have due regard for the guide to safe practice "The boarding and landing of pilots by pilot boat - Code of Practice" published by the British Ports Federation or any document replacing it.

24.4 Gangways and Accommodation Ladders

24.4.1 When provided, gangways and accommodation ladders should be manufactured to a recognised national or international standard and be clearly marked with the manufacturer's name, the model number and the maximum design angle of use and the maximum safe loading by number of persons and by total weight.

24.4.2 A gangway should be carried on a vessel of 30 metres in length and over.

24.2.3 Accommodation ladders should be provided on a vessel of 120 metres in length and over.

24.4.4 Access equipment and immediate approaches to it should be adequately illuminated.

24.4.5 Reference standards include:

BSMA 78:1978 - Gangways (excluding the maximum overall widths specified in table 2);

and

BSMA 89:1980 - Accommodation Ladders.

25 Clean Seas

25.1 Vessels should comply with all the requirements of MARPOL according to the regulations of the Administration.

25.2 Special local requirements may exist in national sea areas, ports and harbours. The attention of owners/operators is drawn to the need to comply with local requirements as appropriate.

26 Manning

26.1 Manning Requirements

26.1.1 All sea going commercial yachts of more than 24 metres Load Line Length but less than 3000 GT should carry qualified Deck and Engineer Officers as required in Annex 5.

26.1.2 All sea going commercial yachts of more than 3000 GT should carry qualified Deck Officers as required by the Merchant Shipping (Certification of Deck Officers) Regulations 1985 as amended.

26.1.3 All sea going commercial yachts of 3000 GT and over should carry qualified Engineer Officers as required by the Merchant Shipping (Certification of Marine Engineer Officers and Licensing of Marine Engine Operators) Regulations 1986 as amended.

26.1.4 The certification of officers referred to in Annex 5 relates to qualifications that have been attained in conjunction with the United Kingdom Department of Transport and/or the Royal Yachting Association and/or the Nautical Institute.

(Officers who have qualifications issued by Administrations and/or Yachting Associations which are recognised by this Administration, will be considered on an individual basis.)

26.1.5 Personnel within the commercial yachting industry who wish to obtain any of the Department of Transport qualifications listed in Annex 5 should refer to Marine Guidance Note MGN.14.(m) - Certificates of Competency for Service on Commercially Operated Yachts and Sail Training Vessels: Arrangements to upgrade RYA qualifications.

26.2 Owners Responsibility

All sea going commercial yachts either of more than 24 metres Load Line length or over, should be safely manned and it is the responsibility of the owner/managing agent to ensure that the master and, where necessary, other members of the crew of a vessel have, in addition to the qualifications required in Annex 5, recent and relevant experience of the type and size of vessel and of the type of operation in which she is engaged.

26.3 Radio Qualifications

Every vessel should carry, as a minimum, one person who holds an appropriate radio operator's certificate which is suitable for the radio installation and equipment on board, and reflects the operating area of the vessel.

26.4 Medical Fitness Certificates

26.4.1 All Officers who are required to hold Department of Transport Certificates of Competency or the equivalent, are required to hold current Medical Fitness Certificates issued by the Department of Transport (Form ENG1) or the MSA (Form ENG1) or by a National Administration in compliance with the requirements of Article 2(a)(iii) of the convention concerning minimum standards in Merchant Ships, Convention 1976 (ILO No 147), under Regulations accepted as equivalent to Medical Examination (Seafarers) Convention 1946 (No 73).

26.4.2 All other Deck and Engineer Officers are required to provide evidence of medical fitness. This evidence should be at least to the standard required for the RYA Commercial Endorsement which requires the satisfactory completion of the MSA form ML5. The holding of a current RYA Commercial Endorsement will be accepted as proof of medical fitness. For those officers not required to hold a Department of Transport Certificate of Competency or RYA Commercial Endorsement, the ML5 may be completed by any registered medical practitioner who must also complete a Certificate in the format specified at Annex 8. It is this Certificate that needs to be produced by the officer concerned as evidence of medical fitness. The ML5 should be retained by the seafarer.

26.5 Basic Sea Survival Course

All deck officers should hold a Basic Sea Survival Course Certificate which is recognised by the Administration.

It is highly recommended that all other members of the vessel's crew should hold Basic Sea Survival Course Certificates.

26.6 First Aid Course

26.6.1 All officers should hold a First Aid At Sea Certificate which is acceptable to the Administration.

26.6.2 All masters on vessels operating more than 60 miles from a safe haven, or on vessels which are operating in an area where no adequate medical facilities are available, should hold a Ship Captain's Medical Training Certificate which is acceptable to the Administration.

26.7 Fire Fighting Course

All officers should hold a Basic Fire Fighting Course Certificate which is acceptable to the Administration.

26.8 Revalidation of Certificates and Licences

26.8.1 All RYA/DTp Yachtmaster Certificates, DOT Certificates of Competency and Licences should be revalidated every five years. To revalidate, the applicant should prove at least one year's service on sea going vessels of more than 24 metres Load Line length or of over 80 GT during the previous five years and be in possession of a valid Medical Fitness Certificate.

26.8.2 Applicants for revalidation who are not able to prove the requisite sea service but are able to demonstrate that during at least half of the 5 year period they have been employed on duties closely associated with the management and operation of one or more of the appropriate types of vessels, may have their Certificates or Licences considered for revalidation.

27 Passengers

No vessel to which the Code applies should carry more than 12 passengers on a voyage or excursion. The following meanings apply:

"*Passenger*" means any person carried in a ship except:

(a) a person employed or engaged in any capacity on board the ship on the business of the ship;

(b) a person on board the ship either in pursuance of the obligation laid upon the master to carry shipwrecked, distressed or other persons, or by reason of any circumstances that neither the master nor the owner nor the charterer (if any) could have prevented; and

(c) a child under one year of age; and

"*a person employed or engaged in any capacity on board the vessel on the business of the vessel*" may reasonably include:

.1 bona-fide members of the crew over the minimum school leaving age (about 16 years) who are properly employed on the operation of the vessel;

.2 person(s) employed by the owner in connection with business interests and providing a service available to all passengers; and

.3 person(s) employed by the owner in relation to social activities on board and providing a service available to all passengers.

The above persons engaged on the business of the vessel should be included in the crew agreement which is required for the vessel.

28 Survey, Certification, Inspection and Maintenance

28.1 General

28.1.1 All ships covered by this Code will have to be surveyed and certificated in accordance with the International Load Line Convention; oil tankers over 150 GT and all other ships over 400 GT under the MARPOL Convention. Cargo ships over 500 GT and all passenger ships undertaking international voyages are required to be surveyed and certificated under the construction and safety equipment requirements of the SOLAS Convention. All ships of over 300 GT and all passenger ships are also required to be surveyed and certificated under the radio requirements of SOLAS. Annex 6 is the list of certificates to be issued.

28.1.2 The underlying principle is that whether a ship is registered in the United Kingdom or in a relevant overseas territory being a Category 1 registry, the same survey standards will apply. Therefore, it has been agreed that this statutory work may be undertaken by surveyors of the Administration or by surveyors appointed by the Administration from Lloyd's Register of Shipping or the British Committees of Bureau Veritas, Det Norske Veritas, Germanischer Lloyd or Registro Italiano Navale or the British Technical Committee of the American Bureau of Shipping and, for safety radio, an appropriate Certifying Authority in relation to radio installations for cargo ships.

28.1.3 A ship on the register to which the International Conventions apply must be surveyed and, if they meet the necessary standards, Convention certificates will be issued to them. All requests for survey and certification must be made to the Administration, the appropriate Classification Society or appropriate Certifying Authority in relation to radio installations to make the required arrangements.

28.2 Initial Survey (including newbuilding commercial yachts)

28.2.1 General

28.2.1.1 A request for survey and certification should be made by the owner(s)/managing agent(s) direct to the Administration or a recognised Classification Society or, in relation to radio installations for cargo ships, an appropriate Certifying Authority.

28.2.1.2 In the case of a commercial yacht located outside the United Kingdom and for which survey and certification is the responsibility of surveyors of the Administration, a Classification Society Surveyor may be appointed to undertake the day-to-day surveys required and to act as a focal point between the shipbuilders/shiprepairers and the Administration.

28.2.1.3 Fees for the surveys conducted by surveyors of the Administration and other authorised organisations will be recovered directly from the owner(s)/managing agent(s) by the Administration/organisations at the prevailing rates.

28.2.2 Load Line Certificate (Yacht to which the Code applies)

28.2.2.1 When a commercial yacht is either in class or under survey to be classed with a Classification Society which is recognised by the Administration as an assigning authority for the load line rules, that Classification Society will be authorised to do the survey and the issue of the International Load Line Certificate (1966).

28.2.2.2 In any other case, the Administration will arrange the survey and issue of the International Load Line Certificate (1966).

28.2.2.3 The Administration will approve stability information for motor yachts up to 100 metres in load line length and sailing yachts of any length.

28.2.2.4 Exemption from the requirements of the Code for load line marking, conditions of assignment and protection of the crew should not be envisaged. Only the Administration has the power to issue a Load Line Exemption Certificate (1966) and it is not policy to exercise this power for commercial yachts.

28.2.3 Certificate of Compliance (Yacht of less than 500 GT)

The Administration will issue a Certificate of Compliance with the Code of Practice. (The form of the Certificate is in Annex 7.)

28.2.4 Cargo Ship Safety Construction Certificate (Yacht of 500 GT or over)

The Classification Societies will survey ships and issue Cargo Ship Safety Construction Certificates. However, the survey of structural fire protection and means of escape on new buildings and ships flagging-in will be undertaken by surveyors appointed by the Administration who will in such cases provide the Societies with partial Declarations of Survey when required.

28.2.5 Cargo Ship Safety Equipment Certificate (Yacht of 500 GT or over)

Surveys of safety equipment on vessels in a port in the United Kingdom, or in the territorial waters thereof will be carried out by a surveyor of the Administration. Otherwise the survey may be carried out by an appropriate Classification Society.

28.2.6 Cargo Ship Safety Radio Certificate (Yacht of 300 GT or over)

The survey of radio equipment in accordance with the issue of a Cargo Ship Radio Certificate will be undertaken by a surveyor appointed by an appropriate Certifying Authority in relation to radio installations for cargo ships.

28.2.7 International Oil Pollution Prevention Certificate (Yacht of 400 GT or over)

The survey for issue of the initial International Oil Pollution Prevention Certificate will be undertaken by surveyors appointed by the Administration.

28.2.8 International Tonnage Certificate (Yacht to which the Code applies)

The Administration or a recognised Classification Society may undertake all surveys for the issue of an International Tonnage Certificate.

28.2.9 Crew accommodation

The approval of crew accommodation will be undertaken by the Administration in compliance with the standards defined in section 21. No certificate is required and compliance will be notified by letter. Any request for exemption from a specified standard should be made to the Administration for a decision.

28.2.10 Safe Manning Document (Yacht of 500 GT or over)

A safe manning document should be kept onboard the yacht at all times, in accordance with the Merchant Shipping (Safe Manning Document) Regulations 1992, SI 1992 No.1564. The Administration will issue the safe manning document.

28.2.11 Exemption from certain safety regulations

If an owner/managing agent seeks an exemption from the application of specific safety regulation(s) formal application must be made to the Administration. The Administration will issue an exemption if and when appropriate.

28.3 Periodical Surveys

28.3.1 Load Line Certificates, Cargo Ship Safety Construction Certificates and Certificates of Compliance (Valid for 5 years in general)

Annual, intermediate and renewal surveys with respect to the Load Line Certificates, the Cargo Ship Safety Construction Certificates and the Certificate of Compliance should be carried out to the satisfaction of the certifying authority.

No extension is permitted to the five year period of validity of these certificates.

28.3.2 Cargo Ship Safety Equipment and Safety Radio Certificates (Valid for 2 years and 1 year respectively)

28.3.2.1 Safety Equipment Certificates and Safety Radio Certificate annual and renewal surveys should be carried out either by the Administration or Parties to the SOLAS Convention at the request of a British Consul, High Commissioner or Governor; or by a Classification Society surveyor appointed by the Administration to act on its behalf; or by an appropriate Certifying Authority in relation to radio installations for cargo ships, except when a ship is in the United Kingdom or the Near Continent when it may be surveyed by the Administration or the Certifying Authority in relation to radio installations.

28.3.2.2 An application to the Administration for an extension to the certificate will be agreed only in cases when it appears proper and reasonable to do so.

28.3.2.3 At least once during a five year period, a surveyor appointed by the Administration will visit the ship to survey its safety equipment and to conduct a general inspection to ensure that standards are being met.

28.4 Use of an Authorised Classification Society

An authorised Classification Society is aware of the extent to which responsibility has been delegated to issue Load Line Certificates and Cargo Ship Safety Construction Certificates. International Conventions give specific discretion to an Administration to either make exemptions or accept equivalent equipment or arrangements. The formal agreement between the Administration and an authorised Classification Society governs the relationship between the two parties.

28.5 Use of a Classification Society Surveyor to act on the behalf of the Administration

An exclusive surveyor from an authorised Classification Society and proposed by the Society may be appointed from time to time to act on behalf of the Administration in cases when it is impracticable for a surveyor of the Administration to make the visit necessary for the survey.

When a Classification Society surveyor is so appointed, actions taken will be under direct instruction of the Administration. The Administration will provide the appointed surveyor with detailed guidance on the scope of survey and report required.

In general, only full-time salaried surveyors who work exclusively for the Society will be appointed by the Administration.

28.6 Accident Investigations

The Administration with which the vessel is registered is obliged to investigate accidents or incidents in accordance with the requirements of International Conventions. Apart from this legal requirement, an Administration investigates such occurrences to demonstrate the effective control and importance they attach to safety at sea.

It is an offence for the vessel's master, skipper or owner not to inform the appropriate authority of a reportable accident shortly after it occurs and to provide details so that an assessment of its seriousness can be made quickly. The Marine Administration or, in the case of the United Kingdom, the Marine Accident Investigation Branch (MAIB), will appoint a suitable Surveyor or Inspector whenever an investigation is required. The Marine Administration or the MAIB, as appropriate, will then receive the Surveyor's or the Inspector's report and will deal with the follow up action.

All serious casualties in accordance with the International Maritime Organisation's (IMO) definition should be reported to IMO through the United Kingdom's MAIB.

ANNEX 1

CATEGORISATION OF RED ENSIGN SHIPPING REGISTERS

1 Background

1.1 Section 18 of the 1995 Merchant Shipping Act contains provisions for regulating the registration of ships in overseas territories by Order in Council. Any Order made to establish different categories of register may include restrictions according to the size and type of ships which may be registered.

1.2 In practice, two categories of register have been recognised:

Category 1 Able to administer and maintain internationally agreed standards, as defined in the relevant International Conventions - SOLAS, STCW, MARPOL, Load Line, Tonnage, and ILO 147; and

Category 2 Generally not permitted to register passenger ships of any size or other ships of over 150 GT.

2 Category 1 Registers

2.1 The overseas territories assigned to Category 1 are:-

Bermuda;
Cayman Islands; and
Isle of Man.

2.2 There is no restriction as to the ships which may be accepted on a register of Category 1. However, there may be local laws that limit registration of certain types of ship.

2.3 The overseas territories listed in 2.1 have in place all the legal requirements, similar to those in force in the United Kingdom, for giving effect to the international conventions listed below:-

International Convention for the Safety of Life at Sea, 1974 as modified by the Protocol of 1978 relating thereto (SOLAS 74/78) as amended.

Convention on the International Regulations for Preventing Collisions at Sea, 1972 (COLREGS 1972)

International Convention on Load Lines 1966 (LL 1966).

International Convention on Tonnage Measurements of Ships, 1969 (TONNAGE 1969).

International Convention for the Prevention of Pollution from Ships, 1973 as modified by the Protocol of 1978 relating thereto (MARPOL 73/78) as amended.

International Convention on Standards of Training, Certification and Watchkeeping for Seafarers, 1978 (STCW 1978) as amended.

International Labour Organisation Merchant Shipping (Minimum Standards) Convention No. 147 (1976) (ILO 147).

2.4 The overseas territories listed in 2.1 are not in a position to give effect to the International Maritime Dangerous Goods Code, 1990 (IMDG Code) as amended. The United Kingdom is the only Red Ensign Administration which can act as a competent authority for this purpose.

2.5 Ships on Category 1 registers are British ships and it is the policy of the United Kingdom and the other respective Governments to ensure that the registers maintain the same standard as the United Kingdom and do not become a haven for substandard ships. It has been agreed that surveyors from the United Kingdom Administration, on request from these registers, may undertake surveys on their behalf in the United Kingdom and, where practicable, in other locations.

2.6 Flag state legislation applies but the survey standards that are applicable to United Kingdom ships also apply to ships on these registers. Therefore, surveyors can use United Kingdom statutory instruments, Instructions to Surveyors and Merchant Shipping Notices for guidance when undertaking surveys.

3 Category 2 Registers

3.1 The effect of Orders in Council (subject to 3.2) is that from a date set in the Orders, Category 2 registers may not accept applications to register from passenger ships or other ships of over 150GT.

3.2 An overseas territory being a Category 2 registry may request re-categorisation to Category 1 which, if acceptable to and agreed by the United Kingdom Secretary of State for Transport, would mean that, in addition to the ships currently on the register, it would be free to accept and retain any other ships of any size or type. Re-categorisation can take place at any time.

3.3 For information, the overseas territories currently assigned to Category 2 are:-

Anguilla;
British Virgin Islands;
Falkland Islands;
Guernsey;
Jersey;
Montserrat;
St. Helena; and
Turks and Caicos Islands.

ANNEX 2

LIST OF REFERENCE DOCUMENTS

REFERENCE SECTION	DOCUMENT
1.1	Merchant Shipping (Categorisation of Registries of Overseas Territories) Order 1992 SI 1736/1992
1.3	International Convention on Load Lines 1966
1.4	"
1.11	Merchant Shipping (Vessels in Commercial Use for Sport or Pleasure) Regulations 1993 SI 1072/1993 as amended
3.1	"
2	Merchant Shipping (Life-Saving Appliances) Regulations 1986 SI 1066/1986 as amended
13.1.2	"
4.4.1.2	SOLAS - Safety of Life at Sea, Chapter II-1
4.5.2	"
7A1.1	"
7B	"
8B	"
9B	"
9B	"
10B	"
14B.2.2	SOLAS - Safety of Life at Sea Chapter, II-2
14B.3	"
15B.2	"
24.2.1	"
13.1.4	Merchant Shipping Notice No. M.1444 Use and Fitting of Retro-Reflective Material on Life-Saving Appliances
14.1.5	BS EN 60529 1992 Specification for Degrees of Protection Provided by Enclosures (IP Code)
14A.8	BS 5852 Part 1 1979 Methods of Test for Assessment of the Ignitability of Upholstered Seating by Smouldering and Flaming Ignition Sources
14B.3.11.8.1	"
14A.8.2	BS 5867 Part 2 1980 Specification for Fabrics for Curtains and Drapes - Flammability Requirements
14B.3.11.8.2	"
14A.8	Consumer Protection, The Furniture and Furnishings (Fire)(Safety) Regulations 1988 Schedule 3
14B.3.11.8.1	"
14A.9	EC Directive 90/396/EEC relating to Appliances Burning Gaseous Fuels
14B.3.12.3	"
14B.3.13.1.3	IMO Resolution MSC.44(65) Standards for Fixed Sprinkler Systems for High Speed Craft

16.2.2	Merchant Shipping (Radio Installations) Regulations 1992 SI 3/1992 as amended
16.7.6	Admiralty List of Radio Signals Volume 3
17	International Regulations for Preventing Collisions at Sea 1972 as amended
19.1	Merchant Shipping (Carriage of Nautical Publications) Rules 1975 SI 700/1975 as amended
22.6.1	Code of Practice for Noise Levels in Ships - Second Edition HMSO 1990
23	Merchant Shipping and Fishing Vessels (Medical Stores) Regulations 1995, SI 1802/1995 as amended
24.3	The Boarding and Landing of Pilots by Pilot Boat - Code of Practice published by the British Ports Federation
26.1.2	Merchant Shipping (Certification of Deck Officers) Regulations 1985 SI 1306/1985 as amended
26.1.3	Merchant Shipping (Certification of Marine Engineer Officers and Licensing of Marine Engine Operators) Regulations 1986 SI 1935/1986 as amended
26.1.5	Marine Guidance Note MGN.14.(m) - Certificates of Competency for Service on Commercially Operated Yachts and Sail Training Vessels: Arrangements to Upgrade RYA Qualifications
26.4.1	Merchant Shipping (Medical Examination) Regulations 1983 SI 808/1983

ANNEX 3

MEMBERS OF THE STEERING COMMITTEE & WORKING GROUP RESPONSIBLE FOR THE CODE

Steering Committee

Marine Safety Agency (Chair)
British Marine Industries Federation
Camper and Nicholsons
Isle of Man Marine Administration
Jon Bannenberg Ltd
Royal Yachting Association
Three Quays Marine Services Ltd

Working Group

Marine Safety Agency (Chair)
American Bureau of Shipping
Association of Sail Training Organisations
British Marine Industries Federation
Bureau Veritas
Burness Corlett & Partners Ltd
Camper and Nicholsons
Hong Kong Government Office
Marine Engineers Certifying Authority Ltd
Nautical Institute
Nigel Burgess Ltd
Professional Yachtmen's Association
Royal Institution of Naval Architects
Royal Yachting Association
Shipbuilders & Shiprepairers National Association
Society of Consulting Marine Engineers and Ship Surveyors
Three Quays Marine Services Ltd
Yacht Brokers Designers and Surveyors Association

ANNEX 4

OPEN FLAME GAS INSTALLATIONS

1 General Information

1.1 Possible dangers arising from the use of liquid petroleum gas (LPG) open flame appliances in the marine environment include fire, explosion and asphyxiation, due to leakage of gas from the installation.

1.2 Consequently, the siting of gas-consuming appliances and storage containers and the provision of adequate ventilation to spaces containing them, is most important.

1.3 It is dangerous to sleep in spaces where gas-consuming open flame appliances are left burning, because of the risk of carbon monoxide poisoning.

1.4 LPG which is heavier than air, when released, may travel some distance whilst seeking the lowest part of a space. Therefore, it is possible for gas to accumulate in relatively inaccessible areas, such as bilges, and diffuse to form an explosive mixture with air, as in the case of petrol vapour.

1.5 A frequent cause of accidents involving LPG installations is the use of unsuitable fittings and improvised "temporary" repairs.

2 Stowage of Gas Containers

2.1 Gas containers should be stowed on the open deck or in an enclosure opening only to the deck or overboard and otherwise gastight, so that any gas which may leak from the containers can disperse overboard.

2.2 In multiple container installations a non-return valve should be placed in the supply line near to the stop valve on each container. If a change-over device is used, it should be provided with non-return valves to isolate any depleted container.

2.3 Where more than one container can supply a system, the system should not be used with a container removed.

2.4 Containers not in use or not being fitted into an installation should have the protecting cap in place over the container valve.

3 Fittings and Pipework

3.1 Solid drawn copper alloy or stainless steel tube with appropriate compression or screwed fittings are recommended for general use for pipework in LPG installations.

3.2 Aluminium or steel tubing, or any material having a low melting point, such as rubber or plastic, should not be used, except as permitted by paragraph 3.3.

3.3 Lengths of flexible piping (if required for flexible connections) should be kept as short as possible and be protected from inadvertent damage. Also, the piping should conform to an appropriate standard.

Proposals for a more extensive use of flexible piping (which conforms to an internationally recognised standard for its application) should be submitted to the Administration for approval on an individual basis.

4 Open Flame Heaters and Gas Refrigerators

4.1 When such appliances are installed, they should be well secured to avoid movement and, preferably, be of a type where the gas flames are isolated in a totally enclosed shield where the air supply and combustion gas outlets are piped to open air.

4.2 In refrigerators, where the burners are fitted with flame arrestor gauze, shielding of the flame may be an optional feature.

4.3 Refrigerators should be fitted with a flame failure device.

4.4 Flue-less heaters should be selected only if fitted with atmosphere-sensitive cut-off devices to shut off the gas supply at a carbon dioxide concentration of not more than 1.5% by volume.

4.5 Heaters of a catalytic type should not be used.

5 Flame Failure Devices

All gas consuming devices should be fitted, where practical, with an automatic gas shut-off device which operates in the event of flame failure.

6 Gas Detection

6.1 Suitable means for detecting the leakage of gas should be provided in any compartment containing a gas-consuming appliance, or in any adjoining space of a compartment into which the gas (more dense than air) may seep.

6.2 Gas detectors should be securely fixed in the lower part of the compartment in the vicinity of the gas-consuming appliance and in other space(s) into which gas may seep.

6.3 Any gas detector should, preferably, be of a type which will be actuated promptly and automatically by the presence of a gas concentration in air of not greater than 0.5% (representing approximately 25% of the lower explosive limit) and should incorporate an audible and a visible alarm.

6.4 Where electrical detection equipment is fitted, it should be certified as being flame-proof or intrinsically safe for the gas being used.

6.5 In all cases, the arrangements should be such that the detection system can be tested frequently whilst the vessel is in service.

7 Emergency Action

7.1 A suitable notice, detailing the action to be taken when an alarm is given by the gas detection system, should be displayed prominently in the vessel.

7.2 The information given should include the following:

7.2.1 The need to be ever alert for gas leakage; and

7.2.2 When leakage is detected or suspected, all gas-consuming appliances should be shut off at the main supply from the container(s) and NO SMOKING should be permitted until it is safe to do so.

7.2.3 **NAKED LIGHTS SHOULD NEVER BE USED AS A MEANS OF LOCATING GAS LEAKS.**

ANNEX 5

MANNING SCALE FOR COMMERCIALLY OPERATED MOTOR YACHTS OVER 24M

(All RYA/DTp Yachtmaster and Coastal Skipper certificates must be Commercially Endorsed.)

AREA	Officer type	VESSEL			
MILES FROM A SAFE HAVEN	Deck or Engineer	> 24m < 200GT	200-500GT	500GT-3000GT Under 3000kW	500GT-3000GT Over 3000kW
Up to 60	Deck Deck	YM Offshore –	5CE or 5CE(Y) Coast Skipper	4CE or 4CE(Y) YM Offshore	4CE or 4CE(Y) YM Offshore
	Engineer Engineer Engineer	AEC* – –	AEC* – –	Class 4(E) AEC* + Shore support	Class 4(E) AEC* + Shore support
Up to 150	Deck Deck	YM Offshore Coast Skipper	5CE or 5CE(Y) Coast Skipper	4CE or 4CE(Y) YM Offshore	4CE or 4CE(Y) YM Offshore
	Engineer Engineer Engineer	AEC* – –	MEOL* – –	Class 4(E) MEOL* + Shore support	Class 4(E) MEOL* + Shore support
Over 150	Deck Deck Deck	YM Ocean YM Offshore –	4CE or 4CE(Y) YM Offshore Coast Skipper	4CE or 4CE(Y) YM Ocean Coast Skipper	4CE or 4CE(Y) YM Ocean Coast Skipper
	Engineer Engineer	SMEOL* AEC*	Class 4(E) AEC*	Class 2(E) Class 3	Class 1 Class 2

In this table the meanings are as follows:-

4CE — DTp Class 4 with Command Endorsement (Merchant Navy) (able to serve to limitations of certificate)

4CE(Y) — DTp Class 4 with Command Endorsement (Limited to yachts)

5CE — DTp Class 5 with Command Endorsement (Merchant Navy) (able to serve to limitations of certificate)

5CE(Y) — DTp Class 5 with Command Endorsement (Limited to yachts)

Coast Skipper — RYA/DTp Coastal Skipper with Commercial Endorsement

YM Offshore — RYA/DTp Yachtmaster Offshore with Commercial Endorsement

YM Ocean — RYA/DTp Yachtmaster Ocean with Commercial Endorsement

AEC — Approved Engine Course

SMEOL — Senior Marine Engine Operator Licence

MEOL — Marine Engine Operator Licence

SHORE SUPPORT — May include breakdown, service and maintenance contracts with marine engineering firms, availability of such facilities etc.

* — Can be dual purpose

Class 4(E) — DTp Marine Engineer Officer Class 4 (Motor) with Service Endorsement

Any vessels powered by steam or gas turbines will be considered on an individual basis.

(It is the overriding responsibility of the Owner/Managing Agent to ensure that the Master and, where necessary, other members of the crew have, in addition to the qualifications required in Annex 5, recent and relevant experience of the type and size of vessel and of the type of operation in which she is engaged.)

MANNING SCALE FOR COMMERCIALLY OPERATED FORE AND AFT RIGGED SAILING YACHTS OVER 24M

(All RYA/DTp Yachtmaster and Coastal Skipper certificates must be Commercially Endorsed and be obtained as SAILING VESSEL qualifications.)

AREA	Officer Type	VESSEL		
MILES FROM A SAFE HAVEN	Deck or Engineer	>24m <200GT	200 - 500GT	500 - 3000GT
Up to 60	Deck Deck Engineer	YM Offshore AEC*	5CE or 5CE(Y) YM Offshore AEC*	4CE or 4CE(Y) YM Offshore Class 4(E)
Up to 150	Deck Deck Engineer	YM Offshore+(c) Coast Skipper AEC*	5CE or 5CE(Y) YM Offshore MEOL*	4CE or 4CE(Y) YM Offshore Class 4(E)
Over 150	Deck Deck Deck Engineer	YM Ocean+(c) YM Offshore MEOL*	4CE or 4CE(Y) YM Offshore Coast Skipper MEOL*	4CE or 4CE(Y) YM Ocean Coast Skipper Class 4(E)

In this table the meanings are as follows:-

4CE	DTp Class 4 with Command Endorsement (Merchant Navy) (able to serve to limitations of certificate)
4CE(Y)	DTp Class 4 with Command Endorsement (Limited to yachts)
5CE	DTp Class 5 with Command Endorsement (Merchant Navy) (able to serve to limitations of certificate)
5CE(Y)	DTp Class 5 with Command Endorsement (Limited to yachts)
Coast Skipper	RYA/DTp Coastal Skipper with Commercial Endorsement
YM Offshore	RYA/DTp Yachtmaster Offshore with Commercial Endorsement
YM Ocean	RYA/DTp Yachtmaster Ocean with Commercial Endorsement
(c)	FOR SAIL TRAINING VESSELS ONLY - The Master of a Sail Training Vessel carrying more than 12 trainees is required to either:- Hold the RYA/DTp Certificate specified in the table and be able to prove at least 50 days satisfactory sea service in a position of responsibility on sail training vessels; or:- Hold at least the appropriate Class 4 or 5 Certificate of Competency with a Command Endorsement.
AEC	Approved Engine Course
SMEOL	Senior Marine Engine Operator Licence
MEOL	Marine Engine Operator Licence
*	Can be dual purpose
Class 4(E)	DTp Marine Engineer Officer Class 4 (Motor) with Service Endorsement

Any vessels powered by steam or gas turbines will be considered on an individual basis.

(It is the overriding responsibility of the Owner/Managing Agent to ensure that the Master and, where necessary, other members of the crew have, in addition to the qualifications required in Annex 5, recent and relevant experience of the type and size of vessel and of the type of operation in which she is engaged.)

MANNING SCALE FOR COMMERCIALLY OPERATED SQUARE RIGGED SAILING VESSELS OVER 24M

(All Yachtmaster and Coastal Skipper certificates must be Commercially Endorsed and be obtained as SAILING VESSEL qualifications.)

AREA	Officer type	VESSEL		
MILES FROM A SAFE HAVEN	Deck or Engineer	>24m <200GT	200 - 500GT	500 - 3000GT
Up to 60	Deck Deck Engineer	YM Offshore+(a)+(c) Coast Skipper AEC*	5CE or 5CE(Y) + (a) 5CE or 5CE(Y) + (b) AEC*	4CE or 4CE(Y) + (a) 5CE or 5CE(Y) + (b) Class 4(E)
Up to 150	Deck Deck Engineer	YM Offshore+(a)+(c) Coast Skipper AEC*	5CE or 5CE(Y) + (a) 5CE or 5CE(Y) + (b) MEOL*	4CE or 4CE(Y) + (a) 5CE or 5CE(Y) + (b) Class 4(E)
Over 150	Deck Deck Deck Engineer	YM Ocean+(a)+(c) YM Offshore+(b) MEOL*	4CE or 4CE(Y) + (a) 4CE or 4CE(Y) + (b) YM Offshore MEOL*	4CE or 4CE(Y) + (a) 4CE or 4CE(Y) + (b) YM Offshore Class 4(E)

A square rigged vessel is a sailing vessel described on her Register as one of the following 5 types of vessel:-	
SHIP	A three, four or five masted vessel with square sails on all masts.
BARQUE	A three, four or five masted vessel with square sails on all masts except the aftermost mast which is fore and aft rigged.
BARQUENTINE	A three, four or five masted vessel fore and aft rigged on all masts except the fore mast which is square rigged.
BRIG	A two masted vessel with square sails on both masts.
BRIGANTINE	A two masted vessel with square sails on the foremast and fore and aft sails on the mainmast.

In the table for Commercially Operated Square Rigged Vessels the meanings are as follows:-

4CE — DTp Class 4 with Command Endorsement - (Merchant Navy) (able to serve to limitations of certificate)

4CE(Y) — DTp Class 4 with Command Endorsement (Limited to Yachts) - (this qualification may be gained by upgrading the RYA/DTp Yachtmaster Ocean with Commercial Endorsement)

5CE — DTp Class 5 with Command Endorsement - (Merchant Navy) (able to serve to limitations of certificate)

5CE(Y)	DTp Class 5 with Command Endorsement (Limited to yachts) - (this qualification may be gained by upgrading the RYA/DTp Yachtmaster Offshore with Commercial Endorsement)
Coast Skipper	RYA/DTp Coastal Skipper with Commercial Endorsement
YM Offshore	RYA/DTp Yachtmaster Offshore with Commercial Endorsement
YM Ocean	RYA/DTp Yachtmaster Ocean with Commercial Endorsement
AEC	Approved Engine Course
SMEOL	Senior Marine Engine Operator Licence
MEOL	Marine Engine Operator Licence
*	Can be dual purpose

(a) The Master of a Square Rigged vessel must, in addition to holding the base certification, have served at least 14 days seatime as a Watchkeeping Officer in the vessel and have been assessed as competent to serve as Master of the vessel by the Owners/Operators under an assessment system approved and monitored by the MSA. The Master shall only serve on the vessel, or specified sister vessel, for which the assessment has been undertaken.

(b) A Watchkeeping Officer of a Square Rigged vessel must, in addition to holding the specified base certification, have served at least 14 days seatime in the vessel in a supernumerary capacity and have been assessed as competent to serve as a Watchkeeping Officer of the vessel by the Owners/Operators under an assessment system approved and monitored by the MSA. A Watchkeeping Officer shall only serve on the vessel, or specified sister vessel, for which the assessment has been undertaken.

Assessment System (a & b) For an Owners'/Operators' Assessment System to be approved by the MSA to permit Officers to serve on specific square-rigged sailing vessels, full details must be submitted of the criteria against which assessment will be made and the process of assessment. Such a system requires owners to demonstrate that the applicants have followed an assessment programme which includes proving knowledge of sailing ship terms and methods of working including the following evolutions:-

Tacking	Setting and stowing sails
Wearing	Reefing
Anchoring	Operating at night
Heaving to	Operating in heavy weather
Coping with Squalls	Effect of knockdowns

The management organisation must be able to demonstrate that they have established that the candidates know how to deal with emergencies and have carried out at least two man overboard evolutions and have demonstrated competence in passage planning in an exercise relating to critical circumstances when the weather pattern is adversely changing and deteriorating

The applicant's performance etc shall be recorded in an appropriate personal log book which should be signed by the Master or the Owner.

(c) FOR SAIL TRAINING VESSELS ONLY - The Master of a Sail Training Vessel carrying more than 12 trainees is required to either:-
Hold the RYA/DTp Certificate specified in the table and be able to prove at least 50 days satisfactory sea service in a position of responsibility on sail training vessels;

or:-

Hold at least the appropriate Class 4 or 5 Certificate of Competency with a Command Endorsement.

Class 4(E)	DTp Marine Engineer Officer Class 4 (Motor) with Service Endorsement.

Any Officer who, in addition to holding the base certification, holds the Nautical Institute Square Rig Sailing Ship Certificate, shall be considered to have met the requirements of paragraphs a and b.

(It is the overriding responsibility of the Owner/Managing Agent to ensure that the Master and, where necessary, other members of the crew have, in addition to the qualifications required in Annex 5, recent and relevant experience of the type and size of vessel and of the type of operation in which she is engaged.)

ANNEX 6

LIST OF CERTIFICATES TO BE ISSUED

Certification	Subject & Convention	United Kingdom Regulations	Survey & Certification Tasked To	Limits	Detail & Remarks
International Tonnage Certificate	Tonnage ITC 69	SI 841/1982	Class		
International Load Line Certificate	Load Line ILLC66	SI 1053/1968	Class	$\geq$ 24 metres	Intact Stability & Subdivision Standard; # - Using equivalent standards of the Code
Safety Construction Certificate	Construction SOLAS 74	SI 1210/1995	Class	$\geq$ 500 GT	Construction
	Fire Protection SOLAS 74	SI 1210/1995	MSA	$\geq$ 500 GT	Structural Fire Protection & Means of Escape
Safety Equipment Certificate	Fire Appliances SOLAS 74	SI 1210/1995	MSA	$\geq$ 500 GT	Fire Appliances
	Life-Saving Appliances SOLAS 74	SI 1210/1995	MSA	$\geq$ 500 GT	Life-Saving Appliances
	Navigation Equipment	SI 1210/1995	MSA	$\geq$ 500 GT	Navigation lights, sounds, signals etc.
Safety Radio Certificate	Radio SOLAS 74	SI 1210/1995	Certifying Authority	$\geq$ 300 GT	
Safe Manning Document	Manning STCW/SOLAS	SI 1564/1992	MSA	$\geq$ 500 GT	
International Oil Pollution Prevention Certificate	Pollution MARPOL	SI 2154/1996	MSA for initial survey only otherwise Class	$\geq$ 400 GT	Pollution prevention equipment
Exemption Certificate	Various	As Applicable to the subject	MSA		Covers exemption from the regulations relative to the subject
Certificate of Compliance	Various	SI 1072/1993	MSA/Class	< 500 GT	Covers aspects surveyed under the Code for which other certificates are not required.

NOTES:
1. THE ADMINISTRATION RETAINS THE RIGHT TO SURVEY AND ISSUE CERTIFICATES FOR ALL OF THE ABOVE ITEMS.
2. REFERENCES TO SI's MENTIONED ABOVE SHOULD BE CONSTRUED AS INCLUDING ANY AMENDMENT TO THOSE SI's WHICH MAY BE MADE FROM TIME TO TIME.

ANNEX 7

CERTIFICATE OF COMPLIANCE

(Crest/Seal of Issuing Authority)

CERTIFICATE OF COMPLIANCE FOR A LARGE CHARTER YACHT

NAME OF VESSEL	OFFICIAL NUMBER	PORT OF REGISTRY	LOAD LINE LENGTH	GROSS TONNAGE

THIS IS TO CERTIFY

1. that the ship has been surveyed in accordance with the Merchant Shipping (Vessels in Commercial Use for Sport or Pleasure) Regulations 1993 (as amended);

2. has been found to comply with the requirements of the Code of Practice for Large Charter Yachts (Motor and Sailing Yachts in Commercial Use for Sport and Pleasure) (as amended);

3. that the total number of persons for which life-saving appliances are provided is ____; and

4. that the hull of the vessel was surveyed on ________.

This certificate is issued under the authority of the Government of the United Kingdom and Northern Ireland. It will remain in force, unless previously cancelled, until the ____________ day of ______ 19____ subject to the vessel, its machinery and equipment being efficiently maintained, annual surveys and manning complying with the Code of Practice, and to the following conditions:-

Issued at ___________________ on the ____ day of ______ 19____

Signed _____________________ Name ________ Date ________

Note: Annual surveys should be carried out within a three month period either side of the anniversary of the date on which the hull was surveyed as recorded at 4. overleaf.

1st Annual Survey | Place ____________________

Official Stamp | Date ____________________

Surveyor ____________________

2nd Annual Survey | Place ____________________

Official Stamp | Date ____________________

Surveyor ____________________

3rd Annual Survey | Place ____________________

Official Stamp | Date ____________________

Surveyor ____________________

4th Annual Survey | Place ____________________

Official Stamp | Date ____________________

Surveyor ____________________

ANNEX 8

CERTIFICATE OF MEDICAL FITNESS

CERTIFICATE OF MEDICAL FITNESS
based upon
MARINE SAFETY AGENCY (MSA) ML5 FORM

This is to certify that:- ....

(Address) ....

....

....

....

....

(Date of Birth) ....

has been examined by me for medical fitness in accordance with the criteria specified by the MSA in the form ML5 (6/94) and I have placed all assessment ticks in Column 2 of Part B with the exception of Section 1.c.(i), 1.c.(ii) and 1.f.
(If any assessment ticks are placed in Column 1, with the exception of Section 1.f., this Certificate must not be awarded and the seafarer should contact the MSA for advice.)

Signed (Medical Practitioner) ....

Name (Block Letters) ....

Address ....

....

....

Date of Examination ....

Certificate valid until ....

(This certificate is valid for a maximum of 5 years for service on vessels operating under the Code of Practice for Large Charter Yachts and Sail Training Vessels.)

****

The holder of this certificate is accepted by the MSA as being medically fit to go to sea in the capacity of uncertificated Deck or Engineer Officer on vessels operating under the Code of Practice for Large Charter Yachts and Sail Training Vessels.

Signature of holder ....

ACKNOWLEDGEMENTS

The Marine Safety Agency would like to thank the following for their kindness and assistance in supplying photographs for this Code:

Camper & Nicholsons Ltd.

The Jubilee Sailing Trust

Tony Castro Ltd.

Yachting Partners International

Printed in the United Kingdom for The Stationery Office
3/97 C15 G3397 10170